CALCULUS LABS
using MATHEMATICA®

ARTHUR G. SPARKS
JOHN W. DAVENPORT
JAMES P. BRASELTON
Georgia Southern University

📖 HarperCollinsCollegePublishers

Sponsoring Editor: George Duda
Project Editor: Carol Zombo
Design Administrator: Jess Schaal
Cover Design: Terri Ellerbach
Production Administrator: Brian Branstetter
Printer and Binder: Patterson Printing
Cover Printer: Patterson Printing

Calculus Labs Using Mathematica®

Library of Congress Cataloging-in-Publication Data
Sparks, Arthur G.
    Calculus labs using Mathematica / Arthur G. Sparks, John W.
Davenport, and James P. Braselton.
        p.   cm.
    Includes index.
    ISBN 0-06-501196-1
    1. Calculus – Data processing. 2. Mathematica (Computer file)
I. Davenport, John W.   II. Braselton, James P.
III. Title.
    QA303.5.D37S63   1993
    515' .0285'5369–dc20
                                        92-45909
                                        CIP

92 93 94 95 9 8 7 6 5 4 3 2 1

Preface

This project updates the calculus curriculum by taking advantage of current technology. By using software with powerful capabilities in numerical computation, symbolic computation and graphics, we can indeed broaden the scope of calculus. Since much of this software is being used in business and industry, the use of current technology in the mathematics curriculum will make the classroom experience more relevant than before.

With the computation power of the computer, methods of approximations and iterative methods can now be done in a meaningful way. With computer graphics, students can now actually see what they could never before even imagine. With the capability to manipulate symbols, the computer can be used to solve very complex problems that here-to-fore would have been unrealistic. With the computer, realistic problems that were impossible to solve before can now be solved routinely. As a result, the students' understanding of calculus will be enhanced and broadened. By using state-of-the-art technology, the classroom experience will be more relevant to the "real world".

OUR APPROACH

To accomplish the preceding, we are using *Mathematica*, a sophisticated software package with powerful capabilities in numerical computation, symbolic manipulation and graphics, to enhance the teaching of calculus and related courses. These courses have been converted into formal laboratory courses. *Mathematica* was chosen because of its overall potential for becoming a truly general purpose mathematics package with very powerful capabilities in numerical computation, symbolic computation and graphics. It has programming capabilities and it is easy to learn and to use effectively. *Mathematica's* "notebook" capability, which allows one to "mix" text and mathematics, has fascinating possibilities for writing mathematical documents. This capability can also be used to great advantage for class demonstrations and lab sessions.

In fact, this approach can supplement the "usual way" courses have been taught. Using the instructor workstations and projection system in the classroom, what was once a passive atmosphere is now more active. In the lab, mathematics is made more exciting to the students by getting them personally involved in experimentation and discovery. The students are not required to understand less than before. They are still expected to handle computations as before. The computer is used for the tedious and complex manipulations. The students are led through a discovery of the principles involved regardless of the topic. Students can be led through concrete and realistic examples without being lost in the tedious and cumbersome calculations. Mathematics is made more real to the students by making the classroom/lab experience more relevant to the "real world". With the aid of the computer, students gain a better understanding of calculus than was available from the traditional approach to these courses.

At Georgia Southern, we have a wide range of students taking calculus. While we use technology to handle more complex problems and provide many applications, we have chosen to put more emphasis on enhancing and broadening the students' understanding of calculus. To this end, our approach in these labs is a guided and structured one in which graphics plays a very important role. These materials are designed to improve on the traditional course and to serve as an ideal supplement for those desiring to begin the use of technology even while using a "traditional" calculus text.

MORE DISTINGUISHING FEATURES

■ THE CONCEPTUAL STRUCTURE OF EACH LAB

Each lab is interactive in nature and thus the students get personally involved in experimentation and discovery.

Space is provided in the manual for students to answer the questions. This provides built-in structure for the written lab report, which is turned in with the student's disk containing the activity for that lab. The written report serves two objectives: (1) it gives the students valuable experience in written communication; and (2) it is easier for the instructors to grade the lab reports.

Some the elements of each lab are as follows:

1. **Introduce new *Mathematica* commands** to be used in the lab. The number of new *Mathematica* commands introduced in a lab is held to a minimum. For these new commands, whether built-in or written, a complete explanation is given. Typically, these commands are illustrated in the examples. In this way,

the students are kept from getting "bogged down" with the software and thus can focus on using software as a tool to solve problems.

2. An **example** (or two) is given with steps guiding the students to the solution. This helps the students "get started" and builds their confidence.

3. **More examples** are given, but students are **given less guidance** and **asked more questions** as the lab proceeds. This forces the students to "look back" at what has already been done in order to proceed to the solution. This provides reinforcement both to the concepts and commands already discussed.

4. **Exercises** for all of the labs are given at the end of the lab manual. By putting the exercises at the end, they will not be torn out with the labs and thus will always be available for assignment. These exercises give the instructor considerable flexibility as to the length and difficulty of the lab assignments.

This format makes the lab manual one which can be used by a wide range of calculus students.

■ FLEXIBILITY

The first lab is designed toward familiarizing, or re-familiarizing as the case may be, the students with the hardware and software being used. This enables students to take any calculus course using this lab manual even if they have not used it in a previous course.

While a two-hour lab works best for our students, the labs can also be used for a one-hour formal lab rather than a two-hour formal lab. The instructor can shorten the lab assignment or have the students finish the longer lab during open lab hours. Similarly, they can be used for occasional formal labs with regular open labs, or just all open labs. The instructor can handle these different situations by his/her lab assignments.

In our present situation, each student has his/her own computer in the lab. However, this is not mandatory. Before our lab was fully equipped, students worked in pairs in the lab and this worked quite well.

Thus, the lab manual can be used for large classes or small ones. Similarly, the manual can be used with computer labs having a large number of computers or a small number. It can be used in formal labs and/or open labs and it can be used in lab situations where students work in pairs or work individually in the lab.

■ PROGRAMMED COMMANDS

Programmed commands are used to great advantage. These commands are written by the authors to achieve certain capabilities not available with a built-in command. The programming aspects are transparent to the students since they do not see the actual definition of the command. They are told what the command does and how it is used. The students are thus able to use the commands to advantage without worrying about the coding of them.

■ TEXTBOOK INDEPENDENT

The labs are written by topic and thus can be used with any textbook.

■ GLOSSARY OF STANDARD MATHEMATICA COMMANDS

A glossary contains the basic standard *Mathematica* commands used in the labs. This provides a quick reference for the students.

■ INDEX OF COMMANDS USED

An index of commands used in the labs is provided. This enables students to look up not only definitions of commands, but examples of them as well.

SUPPLEMENTS AVAILABLE UPON ADOPTION

■ PRELIMINARY FILES (ON DISK)

For each lab, the students will be able to "load" a preliminary file from a disk provided to the instructor. This preliminary file will contain the definitions of the programmed commands along with examples illustrating their use. These commands are then available for use throughout the lab without the students having to understand, or even know, how the commands are coded. The students perform their activity within the preliminary file, thus building it into the electronic report of their lab activity.

■ INSTRUCTOR'S MANUAL (ON DISK)

In the instructor's manual, the code is given for the programmed commands. Answers are provided for the questions posed in the labs and solutions are given for the exercises. Additional hints/warnings are given as appropriate.

A WORD OF CAUTION

While *Mathematica* is very sophisticated and powerful, it has limitations (as does every Computer Algebra System). There are some problems it can solve only with great difficulty. In other instances, it will "solve" the problem but give you an incorrect answer. It is always important to check results for plausibility. Everyone, student and instructor, needs to be aware that every software package has limitations and does not do everything correctly.

ACKNOWLEDGMENTS

We would like to thank colleagues who reviewed our work and provided helpful suggestions: Maurino Bautista, Rochester Institute of Technology; Christopher Cotter, University of Northern Colorado; Philip Crooke, Vanderbilt University; Lillie F. Crowley, Lexington Community College; James Delany, California Polytechnic State University; Martin Forest, Louisiana State University; Leon Hall, University of Missouri-Rolla; LaDawn Haws, California State University-Chico; Roger Higdem, College of Idaho; Linda Host, University of Wisconson-La Crosse; Lewis Lefton, University of New Orleans; Thomas D. Morley, Georgia Institute of Technology; Michael O'Reilly, University of Minnesota; Lester Senechal, Mt. Holyoke College; Rod Smart, University of Wisconsin-Madison; Louis Talman, Metropolitan State College-Denver; Jeremy Teitelbaum, University of Illinois-Chicago; John H. Wadhams, Golden West College; Gerald White, Western Illinois University; Randy Wills, Southeastern Louisiana University and Paul J. Zwier, Calvin College.

We would also like to thank the following: Our colleagues at Georgia Southern University, Martha Abell, Lori Braselton, Yingkang Hu, Lila Roberts, and Cynthia Sikes who provided us with helpful suggestions; Sandra Ellwood, the Department's Administrative Secretary who provided some typing support and put in many hours of proofing the material; Wolfram Research, Inc. who were always willing to answer technical questions about *Mathematica*; and George Duda, Acquisitions Editor-Mathematics at Harper Collins for all his invaluable input and guidance.

The Georgia Southern project is supported by the National Science Foundation's Instrumentation and Laboratory Improvement Program through grant # USE-9052022, the Educational Grant Program of Wolfram Research, Inc., and an equipment grant from Apple Computer, Inc..

Table of Contents

CALCULUS LABS USING *MATHEMATICA*

Introduction to Computing with Mathematica

NAME(S):   INSTRUCTOR:
CLASS: DATE:

■ INTRODUCTION

The purpose of this lab is to introduce you to computing with *Mathematica* at a basic level emphasizing the commands that will be used throughout **Calculus Labs Using *Mathematica***.

Although the mathematics here is elementary, you will have to copy and enter commands correctly, interpret your results, and become proficient in using the most basic *Mathematica* commands.

■ NEW *MATHEMATICA* COMMANDS

At the end of this investigation, you will be able to:
1. define and evaluate functions;
2. use the commands **Plot** to graph functions and **Show** to display several graphs simultaneously;
3. use the commands **Solve**, **NRoots**, and **FindRoot** to both solve and numerically approximate solutions of equations;
4. use the commands **Expand** and **Together** to simplify algebraic expressions;
5. use the commands **Table**, **TableForm**, and **Map** to create tables of numbers;
6. use the command **/.** to evaluate functions and expressions;
7. use the command **N** to compute numerical approximations of numbers; and
8. use the command **?** to obtain information about *Mathematica* commands.

NOTE. The capital letter(s) in the built-in *Mathematica* commands is(are) absolutely necessary. If, in a built-in command, you do not capitalize a letter when it should be, *Mathematica* does **not** recognize that command. This means that you should be extremely careful when typing in these commands.

■ GRAPHING FUNCTIONS

Functions are graphed with the **Plot** command. **Plot[f[x],{x,a,b}]** graphs the function f on the interval [a,b]. In each of the following examples, carefully type and **Enter** each of the specified commands. (To **Enter** a command, put the cursor in the cell containing the command and hit the **Enter** key, **not** the Return key.) In each case, neatly sketch the result in the space provided.

EXAMPLE 1. Let $y = \sqrt{x}$.

❏ **1.1.** Graph this function for $0 \leq x \leq 4$ and name the output **graph1** by typing and **Entering** the command **graph1=Plot[Sqrt[x],{x,0,4}]**

In addition, graphs may be displayed in various levels of gray or, if using a color monitor, in different colors by taking advantage of the option **PlotStyle->GrayLevel[r]** or the option **PlotStyle->RGBColor[r,g,b]**, where r, g, and b are numbers between 0 and 1. (The **->** is obtained by hitting - and then >.)

EXAMPLE 2. Let $y = 2\sqrt[3]{x} = 2x^{\frac{1}{3}}$.

❏ **2.1.** Graph this function for $-4 \leq x \leq 4$ and name the result **graph2** by typing and **Entering** the command: **graph2=Plot[2x^(1/3),{x,-4,4},PlotStyle->GrayLevel[.1]]**

The **Plot** command is very powerful. As demonstrated above, graphs can be displayed in various shades of gray. In addition, several functions can be graphed simultaneously.

EXAMPLE 3. Consider the functions $y = \frac{1}{4}x^2$ and $y = \sin(x)$.

❏ **3.1.** To graph both of these functions on the interval [-4,4] and name the result **graph3**, type and **Enter**:

graph3=Plot[{Sin[x],1/4x^2},{x,-4,4},PlotStyle->{GrayLevel[.3],GrayLevel[.4]}]

Another way to graph several functions on the same axes is to create each graph with the command **Plot** and then use the command **Show** to display them simultaneously. **Show[graph1,graph2]** displays both graphs simultaneously. The option **AspectRatio->r** makes **r** the height-to-width ratio for the plot.

EXAMPLE 4.   Display **graph1**, **graph2**, and **graph3** created above.

❏ **4.1.** Type and **Enter**:

Show[graph1,graph2,graph3,AspectRatio->1]

▣ EVALUATING EXPRESSIONS AND SOLVING EQUATIONS

Mathematica can usually solve equations in which explicit solutions can be found with the command **Solve**. If f(x)=g(x) represents an equation in variable x, the command **Solve[f[x]==g[x]]** attempts to solve the equation f(x)=g(x) for x. If *Mathematica* cannot solve the equation, it says so. Notice that with *Mathematica*, a double equals sign ("**==**")is used to separate the left and right-hand side of equations.

EXAMPLE 5. Consider the equation $\dfrac{1}{2}x = \dfrac{1}{4}x^2$.

❏ **5.1.** To solve this equation for x, type and **Enter** the command

Solve[1/2x==1/4x^2]

The result means that the values of x that satisfy the equation $\dfrac{1}{2}x = \dfrac{1}{4}x^2$ are 0 and 2.

❏ **5.2.** Find the value of $\dfrac{1}{2}x$ when x has value 2 by typing and **Enter**ing:

1/2x /.x->2

NOTE. **expression/.x->2** means replace all values of x in **expression** by the value 2.

❏ **5.3.** The graphs of $\dfrac{1}{2}x$ and $\dfrac{1}{4}x^2$ intersect at the points (,   ) and
(,   ).

EXAMPLE 6. Consider the equation $2x^{\frac{1}{3}} = \frac{1}{4}x^2$.

❏ **6.1.** Visualize the solution by **Enter**ing the command
Plot[{2 x^(1/3),1/4 x^2},{x,-2,4}] and sketching the graphs.

❏ **6.2.** Solve the equation for x by typing and **Enter**ing:

Solve[2x^(1/3)==1/4 x^2]

The result means that the only real values of x that satisfy the equation $2x^{\frac{1}{3}} = \frac{1}{4}x^2$

are 0 and $2 \times 2^{4/5}$. The other four solutions, each containing the symbol **I** are
non - real, which we will ignore.

❏ **6.3.** Compute the value of $2x^{1/3}$ and $\frac{1}{4}x^2$ when x has value $2 \times 2^{4/5}$ by

carefully typing and **Enter**ing the following two commands.

1/4 (2 2^(4/5))^2

1/4 x^2 /. x->2 2^(4/5)

❏ **6.4.** The graphs of $2x^{1/3}$ and $\frac{1}{4}x^2$ intersect at the points (,   ) and

(,   ).

▣ DEFINING FUNCTIONS

Even though *Mathematica* has hundreds of built-in commands and functions, users frequently need to define their own functions. Users may define functions in a variety of ways. At this point, however, we will define functions of a single variable in the simplest way possible. Since every built-in *Mathematica* function and command begins with a capital letter, **every user-defined function, command, and expression in this text will be defined using lower case letters**. This way, the possibility of conflicting with a built-in *Mathematica* function (or command) is completely eliminated. Also, since definitions of functions are frequently modified, we introduce the **Clear** command.

Clear[expression] clears all definitions of the object **expression**. We are certain to avoid any ambiguity when we create a new definition of a function if we always clear all prior definitions of the function and use lower-case letters. In addition, the argument of a function must always be enclosed in square brackets and an underline "_" must be placed after the argument on the left-hand side of the equals sign in the definition of the function.

NOTE. The underline is obtained by holding down the Shift key and hitting the - key. (i.e. Shift-)

EXAMPLE 7. Let $f(x) = \dfrac{3x - x^3}{x^2 + 2}$.

❑ **7.1.** Define f by carefully typing and **Enter**ing the following commands:

Clear[f]; f[x_]=(3x-x^3)/(x^2+2)

NOTE. This is the equivalent of:
Clear[f]
f[x_]=(3x^3)/(x^2+2)
Check the output to be sure you entered the definition of f correctly. (**Enter f[x]** if necessary.)
(When you evaluate functions, you must **always** enclose the argument in square brackets.)

NOTE. If you re-start a session, you must re-execute the definitions of the functions you will be using.

❑ **7.2.** Compute f(4), f(-2), and f(a) by typing and **Enter**ing each command:

f[4]

f[-2]

f[a]

Another way to compute **f[number]** is to first compute **f[x]** and then replace **x** by **number**.

❑ **7.3.** Type and **Enter** the following command to first compute f(x) and then replace x by 4. Notice that the same result is obtained by entering the command f[4].

f[x] /. x->4

Combinations of functions and expressions may be manipulated as well. Sometimes *Mathematica* displays results of algebraic manipulation as complex fractions. Complex fractions may be displayed as a single fraction with the command **Together**. In general, the command **Together[expression]** displays **expression** as a fraction with a single denominator.

❑ **7.4.** Compute $\dfrac{f(a + h) - f(a)}{h}$ and then write the result as a single fraction by carefully typing and **Enter**ing the following two commands.

(f[a+h]-f[a])/h

Together[(f[a+h]-f[a])/h]

In some instances, numerical approximations of solutions of equations are more meaningful than the exact value of the solutions. Numerical approximations of solutions of polynomial equations can be obtained with the command **NRoots**. If f(x)=g(x) represents a polynomial equation the command **NRoots[f[x]==g[x],x]** yields numerical approximations of the solutions of the equation f(x)=g(x). Notice that as with the **Solve** command, a double equals sign separates the left and right-hand side of the equation.

Sometimes solutions of equations involve complex numbers. The *Mathematica* symbol **I** denotes the imaginary number $\sqrt{-1}$. In beginning calculus of real variables, you should ignore *Mathematica* output that contains the symbol **I**.

EXAMPLE 8. Let $g(x) = -1 + 7x - x^2 - 9x^3 + 5x^4$.

❑ **8.1.** Define g by carefully typing and **Enter**ing the following commands:
Clear[g]; g[x_]=-1+7x-x^2-9x^3+5x^4

❑ **8.2.** Graph g(x) on the interval [-1.5,1].

❑ **8.3.** After typing and **Enter**ing the following two commands, compare the results. Which has more meaning to you?

Solve[g[x]==0]

NRoots[g[x]==0,x]

❑ **8.4.** Approximate the real values of x that satisfy the equation g(x)=5 by **Enter**ing the following command:

NRoots[g[x]==5,x]

◼ **CREATING TABLES OF NUMBERS**

Tables and lists of numbers, ordered pairs and other expressions are created with the command **Table** and displayed in row and column form with the command **TableForm**.

EXAMPLE 9. Use *Mathematica* to create a table of numbers $\left\{n, n^2, n^3, n^4\right\}$ for

$n = 1, 2, 3, 4, 5$ and display the resulting table in row and column form by typing and

Entering the following command:

```
Table[{n,n^2,n^3,n^4},{n,1,5}]  //  TableForm
```

The exact same results are obtained by typing and **Entering**:

```
TableForm[Table[{n,n^2,n^3,n^4},{n,1,5}]]
```

NOTE. In the above command, {n,1,5} means n assumes the integer values 1,2,3,4, and 5.

❏ **9.1.** Identify the first and last elements of the above table.

Since tables are *Mathematica* objects, they can be named for later use.

EXAMPLE 10. To create a table of numbers $t\dfrac{\pi}{6}$ for $t = 0, 1, 2, \ldots, 11, 12$ and name the

resulting table **vals**, type and **Enter** the following command:

```
Clear[vals];  vals=Table[t Pi/6,{t,0,12}]
```

❏ **10.1.** The ith element of **vals** is obtained with the command **vals[[i]]**. Display the first, fourth, tenth and last elements of **vals**.

If **list** $= \left\{a_1, a_2, \ldots, a_{n-1}, a_n\right\}$ is a list of *Mathematica* objects, then the sum

$a_1 + a_2 + \ldots, a_{n-1} + a_n$ is computed with the command **Apply[Plus, list]**

❏ **10.2.** Compute the sum of all the numbers in **vals** by typing and **Entering** the following command:

```
Apply[Plus, vals]
```

If **f** is a function and **list** $= \left\{a_1, a_2, \ldots, a_{n-1}, a_n\right\}$ is a list of *Mathematica* objects

in the domain of **f**, then the list of numbers $\left\{f(a_1), f(a_2), \ldots, f(a_{n-1}), f(a_n)\right\}$ is

computed with the command **Map[f,list]**

❏ **10.3.** Compute the sine of each number in the list **vals** by typing and **Entering** the following command:

```
Map[Sin,vals]
```

A function f is **listable** means that if **list** $= \{a_1, a_2, \ldots, a_{n-1}, a_n\}$ is a list of

Mathematica objects, then **Map[f,list]** and **f[list]** produce the same result. Many built-in *Mathematica* functions are listable. Hence, the following command produces the same result as in **10.3**.

Sin[vals]

▣ OBTAINING INFORMATION ABOUT *MATHEMATICA* COMMANDS

The previous examples have illustrated the commands **Plot, Solve, NRoots,** /., **Together, Table, TableForm, Map** and **N**. In general, information and syntax about any *Mathematica* command may be obtained using the command **?**

> **EXAMPLE 11.** Obtain information about the command **Expand** by carefully typing and **Entering:**

?Expand

❑ **11.1.** Predict the result of entering the command **Expand[(3x-4y)^2]**

❑ **11. 2.** Use **Expand** to compute $\left(2x^2 - 3y^3\right)^3$.

Functions and Limits of Functions

NAME(S):   **INSTRUCTOR:**
CLASS: **DATE:**

▪ NEW *MATHEMATICA* COMMANDS AND TERMINOLOGY

A **deleted neighborhood** of a point x=a is any open interval that contains the point x=a with that point excluded. An example of a deleted neighborhood of x=1 is the set of points (.9,1) union (1,1.1) (i.e. the open interval (.9,1.1) with the point x=1 removed). This is one means of referring to all the points that are "close to but not equal to a".

The command s=**delnbd**[{a,b},pt,n] can be used to create a set s of n random numbers between a and b, none of which are equal to the number x=pt. Thus **delnbd** returns a list of n random numbers from the deleted neighborhood (a,b)-{pt} of pt. (This command has been created for use in this lab and is not a standard *Mathematica* command. Your instructor will make it available to you.)

Suppose we are given a finite set s of points and a function f, and we want to find the maximum value of f over s and the minimum value of f over s. We can accomplish this by executing the following commands.

```
f[x_]:=Sin[x]      (* Define the function f(x) *)
fs=Map[f,s]        (* Apply f to each point in the set s *)
maxf=Max[fs]       (* Locate the maximum value *)
minf=Min[fs]       (* Locate the minimum value*)
```

Mathematica has a command that attempts to calculate $\lim_{x \to a} f(x)$. The format of this command is

Limit[function,variable->number] where -> is obtained by hitting - and then >.

Thus, to find $\lim_{x \to 2} x^2$, just **Enter** the command **Limit[x^2, x − > 2]**

To find $\lim_{x \to \infty} \dfrac{1}{x-1}$, just **Enter** the command **Limit[1 / (x−1), x − > Infinity]**

NOTE. *Mathematica* can handle infinite limits.

▪ AN INTUITIVE APPROACH TO LIMITS

What happens to the function values of f in a deleted neighborhood of x = a? Do they tend to cluster around a fixed point L? If so, then we say that $\lim_{x \to a} f(x) = L$. If not, then $\lim_{x \to a} f(x)$ DNE (does not exist).

9

Let's see how we can use *Mathematica* to get an intuitive feel as to whether the limit of a particular function does or does not exist. There are two methods that we can use to do this.

METHOD 1. Determine the range of fluctuation of f over each of the following deleted neighborhoods of a by doing the following. (By the range of fluctuation we mean the difference between the maximum value of f over the window and the minimum value of f over the window).

1. Plot f(x) over a symmetric interval about x=a.
2. Use the graph of y = f(x) to determine the largest value of f over this interval.
   NOTE: We can estimate the maximum and minimum value by doing the following:
   a. Select the graph of f by clicking on it.
   b. Hold the apple key down while pressing the mouse key. Note that the cursor changes to a cross-hair. (If you are not using a Macintosh, there should be an equivalent key on your computer. In this event, check with your instructor.)
   c. Place the cursor at a point on the graph. Then, the coordinates for that point will appear in the lower left-hand corner of the screen if you are using a Macintosh.
3. Use the graph of f to determine the smallest value of f over this interval.
4. Find the difference between the largest value and the smallest value.

METHOD 2. The second method also involves looking at smaller and smaller intervals about x=a. For each interval,
1. Generate 100 random values of x that lie in the deleted neighborhood.
2. Find the largest value of f over this set of points.
3. Find the smallest value of f over this set of points.
4. Find the difference between the largest value and the smallest value.

NOTE. *Mathematica* draws the graph of a function by plotting a finite number of points and then connecting them. *Mathematica* will pick more points to plot in those intervals for which it is determined that the curve is changing directions a lot. Is it possible that the graph as depicted by *Mathematica* using this approach would not be a completely accurate picture?

You need to remember that both of these methods only yield approximations to the maximum and minimum values of f on an interval. One of our objectives in calculus is to determine a precise method for finding the true maximum and minimum values of a function over an interval.

EXAMPLE 1. Let $g(x) = (1+x)^{\frac{1}{x}}$ and $a = 0$.

❏ **1.1.** Using a symmetric interval about x=0 of length .2, perform the four steps in METHOD 1. Indicate your results.

❑ **1.2.** Using a symmetric interval about x=0 of length .02, perform the four steps in METHOD 1. Indicate your results.

❑ **1.3.** Using a symmetric interval about x=0 of length .0002, perform the four steps in METHOD 1. Indicate your results.

❑ **1.4.** Using a symmetric interval about x=0 of length .2, perform the four steps in METHOD 2. Indicate your results.

❑ **1.5.** Using a symmetric interval about x=0 of length .02, perform the four steps in METHOD 2. Indicate your results.

❑ **1.6.** Using a symmetric interval about x=0 of length .0002, perform the four steps in METHOD 2. Indicate your results.

❏ **1.7.** Looking at the above results, would you say that the limit of g(x) at x=0 exists? If so, what do you think the limiting value would be?

> **EXAMPLE 2**. Let $h(x)=\sin\left(\dfrac{1}{x-1}\right)$ and $a=1$.

❏ **2.1.** Using a symmetric interval about x=1 of length .2, perform the four steps in METHOD 1. Indicate your results.

❏ **2.2.** Using a symmetric interval about x=1 of length .02, perform the four steps in METHOD 1. Indicate your results.

❏ **2.3.** Using a symmetric interval about x=1 of length .0002, perform the four steps in METHOD 1. Indicate your results.

❏ **2.4.** Using a symmetric interval about x=1 of length .2, perform the four steps in METHOD 2. Indicate your results.

❏ **2.5.** Using a symmetric interval about x=1 of length .02, perform the four steps in METHOD 2. Indicate your results.

❏ **2.6.** Using a symmetric interval about x=1 of length .0002, perform the four steps in METHOD 2. Indicate your results.

❏ **2.7.** Looking at the above results, would you say that the limit of h(x) at x=1 exists? If so, what do you think the limiting value would be?

▣ COMPUTING LIMITS USING THE LIMIT COMMAND

(A caveat: **In unusual cases, *Mathematica* will compute an incorrect answer!!** Therefore, you will have to ask yourself if the answer given by *Mathematica* makes sense.)

> **EXAMPLE 3.** Let $f(x) = \dfrac{x^3 + x^2 - x - 1}{x^3 + 2x^2 + 2x + 1}$

❏ **3.1.** Find, if it exists, $\underset{x \to -1}{\text{Lim}}\ f(x)$

HINT. What is $x^3 + x^2 - x - 1$ evaluated at $x = -1$?

What is $x^3 + 2x^2 + 2x + 1$ evaluated at $x = -1$?

> **EXAMPLE 4.** Let $g(x) = \dfrac{\sin(x-4)}{\sqrt{x}-2}$.

❏ **4.1.** Plot g(x) over an interval containing x=4.

❑ **4.2.** Find, if it exists, $\underset{x \to 4}{\text{Lim}}\, g(x)$.

EXAMPLE 5. Let $h(x) = \dfrac{5x}{\sqrt{x^2 + 4x + 5}}$

❑ **5.1.** Find, if it exists, $\underset{x \to \infty}{\text{Lim}}\, h(x)$

❑ **5.2.** Plot h(x) over the interval (200,10000).

❑ **5.3.** Does this graph "correspond" to your answer in 5.1? Explain.

❑ **5.4.** Plot h(x) over the interval (-10000,-200).

❑ **5.5.** Find, if it exists, $\underset{x \to -\infty}{\text{Lim}}\, h(x)$.

❑ **5.6.** Does your answer "correspond" to the graph in 5.4? Explain.

Continuity

NAME(S):
CLASS:

INSTRUCTOR:
DATE:

▪ INTRODUCTION

Suppose we want to draw a "curve" on a piece of paper. If we put the pencil point down on the paper at one point and draw a curve to another point without removing the pencil point from the paper, then we say that, intuitively, the curve is continuous. One of the important questions about a function is whether or not the function is continuous at a particular point. Loosely speaking, a function is continuous at x=a if it is true that the function values f(x) are "close" to f(a) whenever the x value is

"close" to the point a. Formally, f is continuous at $x = a$ if $\lim_{x \to a} f(x) = f(a)$. Recall that this is true provided both $\lim_{x \to a^+} f(x) = f(a)$ and $\lim_{x \to a^-} f(x) = f(a)$. In this lab we will examine the concept of a

function being continuous at a point and then, more generally, on an interval. To help us better understand this concept, we will also examine ways that a function can be discontinuous (i.e. not continuous) at a point.

▪ NEW *MATHEMATICA* COMMANDS AND PROCEDURES

One command for iterating a given process is the **Do** command. The general format for the **Do** command is **Do[expression,{n,nmin,nmax,nincrement}]**. For example, to plot the functions y = n*x^2 as n ranges from -.5 to .5 changing by .11 each time, we use the following command.

Do[Plot[n x^2, {x,-2,2}, PlotRange->{-1,1}, PlotLabel->n, {n,-.5,.5,.11}]

Other commands being used for the first time are listed below with a brief description of what each command does. The form of each command is the form used in this lab. You need to keep in mind that some of these commands may also have other forms.

1. **Graphics[list]** creates a two-dimensional graphics object.
2. **InterpolatingPolynomial[list of points]** is the polynomial of least degree passing through the points.
3. **linearfit[list of points,m,n]** joins each successive pair of points in the list, from point m to point n, with a line segment and then displays the graph. (This is not a standard *Mathematica* command. We have created this command to use in this lab. Your instructor will make it available to you.)
4. **ListPlot[list of points]** plots the list of points.
5. **Point[{x,y}]** represents a point at position (x,y).
6. **PointSize[s]** gives points a size s measured as a fraction of the width of the whole plot.

7. Show[graphics] displays the graphics object **graphics**.

To animate graphics: (1) Use the cursor to select the series of graphic cells that you want to animate; and (2) Move the cursor to **Graph** in the menu and select **Animate Selected Graphics**.

■ INFINITE BREAKS

> **EXAMPLE 1.** Define $f(x) = \dfrac{1}{x^2}$ for $x \neq 0$ and $f(0) = 1$ by executing the following commands.

   Clear[f]
   f[x_]:=1/x^2 /; x>0
   f[x_]:=1 /; x==0
   f[x_]:=1/x^2 /; x<0

To plot f on the interval [-1,1], type and **Enter** the command **Plot[f[x],{x,-1,1}]**.

❏ **1.1.** Is f continuous at x=0? (That is, when x is close to 0 (but x≠0) is f(x) close to f(0)=1?) If it is not continuous at x=0, why not? (Justify your answer.)
HINT. Plot f[x] over smaller and smaller intervals, such as [-.01,.01], [-.0001,.0001], etc. about x=0. Apply the definition of continuity given in the INTRODUCTION.

❏ **1.2.** If the value of f at x=0 were changed, would this change your answer? If no, why not? If yes, what value could you choose for f at x=0 that would change the function to make it continuous there?

■ JUMP BEHAVIOR

> **EXAMPLE 2.** Define $g(x) = \begin{cases} cx + 1, x \geq 1 \\ 2x^2, x < 1 \end{cases}$ by executing the following commands .

   Clear[g,c]
   g[x_]:= c x + 1 /; x>=1
   g[x_]:= 2 x^2 /; x<1

What does the graph of g look like for different choices of c? Execute the following commands and then animate the set of pictures obtained.

16

```
pointa=Graphics[ PointSize[.02], Point[{1,2}], Point[{1,c+1}] ]
Do[ pfl =Plot[ g[x],{x,-1,1}, PlotLabel->{"c=", c}, DisplayFunction->Identity ];
   pfr=Plot[ g[x],{x,1,4}, PlotLabel->{"c=", c}, DisplayFunction->Identity ];
   Show[ pointa, pfl, pfr, DisplayFunction->$DisplayFunction ],
{c,0,2,.251} ]
```

❏ **2.1** . What is $\lim\limits_{x\to 1^+} g(x)$? (the right - hand limit at x =1) Calculate this limit by hand and

using *Mathematica*.

❏ **2. 2.** What is $\lim\limits_{x\to 1^-} g(x)$? (the left - hand limit at x =1)

HINT. One or both of these answers is expressed in terms of c.

❏ **2.3.** What value of c will result in g(x) being continuous at x=1? Why?

■ RAPID OSCILLATIONS

> **EXAMPLE 3.** Define $f(x)=\sin\left(\dfrac{1}{x}\right), x\neq 0$ and $f(0)=1$.

❏ **3.1.** Is f continuous at x=0? Justify your answer. (HINT. Try to determine whether or not the right-hand limit of f exists. "Zoom in" on the graph of f to the immediate right of x=0 by plotting f over small intervals like [.01,.02], [.001,.0012], etc.)

❏ **3.2.** How would you describe the behavior of f in a neighborhood of x=0? Be as precise as possible.

❏ **3.3.** Is it possible to define f at x=0 in such a way as to make f continuous at x=0? (Justify your answer.)

EXAMPLE 4. Define $g(x) = \begin{cases} 1 & ,x = 0 \\ x\sin\left(\dfrac{1}{x}\right) & ,x \neq 0 \end{cases}$

❑ **4.1.** Is g continuous at x=0? (Justify your answer.)

❑ **4.2.** What is the basic difference in the behavior of f(x) (in EXAMPLE 3) and g(x)?

❑ **4.3.** If g is not continuous at x=0, then is the discontinuity a removable one? If that is the case, how would you define g at x=0 to make it continuous there?

▣ INTERMEDIATE VALUE THEOREM

If f is continuous on a closed interval [a,b] and c is any number between f(a) and f(b), inclusive, then there is at least one number x in the interval [a,b] such that f(x)=c.

A consequence of this is the following :

If f is continuous on a closed interval [a,b], and if f(a) and f(b) have opposite signs, then there is at least one number x in (a,b) such that f(x)=0. (i.e. f has a zero in (a,b).)

NOTE. The Intermediate Value Theorem says that if f is continuous on [a,b], then f assumes all values between f(a) and f(b). Later, you will find that f assumes both a minimum value minf and a maximum value maxf on [a,b]. One can then show that f([a,b]) = [minf,maxf]. Thus the image of a bounded closed interval under a continuous mapping is a bounded closed interval.

EXAMPLE 5. Define $f(x) = x^3 - x - 1 - 3\sin(x)$ on [-1, 2].

Since polynomials and the sine function are continuous, f is continuous.

❑ **5.1.** By evaluation, find the value of f at each of the endpoints.

f(-1)=

f(2)=

18

❑ **5.2.** Try using the **Solve** command to find x in [-1,2] such that f(x)=2.
NOTE: If the **Solve** command does not work satisfactorily, go to **5.3** for an alternate way of obtaining the answer.

x=

❑ **5.3.** Plot f(x)-2 on [-1,2] and approximate a zero **zval** of f(x)-2 in[-1,2]. Note that f(**zval**)-2 approximates 0 and thus f(**zval**) approximates 2 . Thus, **zval** is a good starting point to find the answer to **5.2.** Now, use **FindRoot** with **zval** as your first approximation. (NOTE: The general form is **FindRoot[g[x]==h[x],{x,approx}]**, where **approx** is your first approximation to a solution of g(x)=h(x)). **Enter** the command **FindRoot[f[x]==2,{x,zval}]** to obtain the answer to the question in **5.2.** (**Before** you execute the command, insert your value for **zval**).

zval=
answer=

❑ **5.4.** Verify that your answer is indeed "correct". (Recall that **FindRoot** will give a numerical approximation!)

❑ **5.5.** Find an interval that contains a zero of f and then find such a zero.]

interval=
zero=

■ SPLINES

In many applications, a set of data points is determined and it is desired to fit a "smooth curve" to the points. From the draftsman's point of view, one could use a French curve or a spline (flexible rod) on the drafting table, but this approach is very subjective. From the mathematical viewpoint, if the set of data points is a function, then we want to find a smooth function s(x) which interpolates the data (i.e. the given data points will lie on the graph of s(x)). One way of finding s(x) is for s(x) to be the interpolating polynomial of lowest degree (i.e. s(x) is the polynomial of lowest degree whose graph passes through the given data points). Note that s(x) is smooth and continuous. It can be shown that the degree of the interpolating polynomial may be as large as n-1 for n data points. This is not desirable since polynomials of high degree tend to oscillate wildly and hence do not give a satisfactory fit from an applications standpoint. To illustrate this, consider the following example.

> **EXAMPLE 6.** Measurements have been taken and we have seventeen data points from the cross section of part of a car door. The points are $(0,2.51),(1,3.3),(2,4.04),(3,4.7),(4,5.22),$ $(5,5.54),(6.1,5.8),(6.3,5.55),(6.5,5.44),(6.7,5.4),(7,5.4),(8,5.57),(9,5.7),(12,5.84),$ $(14,5.75),(16,5.48)$ and $(18,4.9)$. We want to fit a smooth curve to the points..

Let's graph the data points and the interpolating polynomial. First, put the data points in a table by **Enter**ing the following.

tbl={ {0,2.51}, {1,3.3}, {2,4.04}, {3,4.7}, {4,5.22}, {5,5.54}, {6.1,5.8},
{6.3,5.55}, {6.5,5.44}, {6.7,5.4}, {7,5.4}, {8,5.57}, {9,5.7},
{12,5.84}, {14,5.75}, {16,5.48}, {18,4.9} }

❏ **6.1.** Indicate the result of **Entering ListPlot[tbl,AspectRatio->.3]**

Now let's find the interpolating polynomial and see if it will be acceptable for this design.

❏ **6.2.** Find the interpolating polynomial by **Enter**ing the command
poly=InterpolatingPolynomial[tbl,x]

Note that the degree of the interpolating polynomial is 16. Now let's plot it on [0,18]. By **Enter**ing **Plot[poly,{x,0,18}]** we obtain

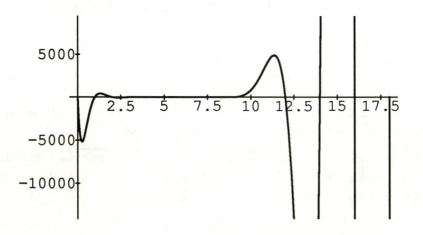

Note the wild oscillations and the large values that the function assumes. Now let's restrict the range and replot it.

By **Enter**ing the command **graph1=Plot[poly,{x,0,18},PlotRange->{-50,50}]** we obtain the following:

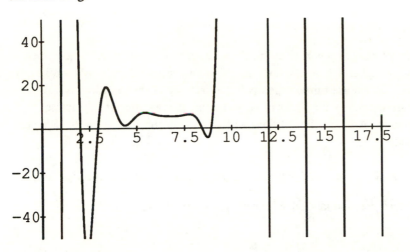

Now let's highlight the data points on the graph of the interpolating polynomial. Before you complete the following, make sure you have **Enter**ed the commands above which defined **tbl**, **poly**, and **graph1**.

❏ **6.2.** Indicate the result of **Enter**ing   **ptbl=Map[Point,tbl]**
graph2=Graphics[{PointSize[.015],ptbl}]
Show[graph1,graph2]

This clearly shows that interpolating polynomials are not satisfactory in design work. An approach that would eliminate the wild oscillations would be to just connect each pair of successive points with a line segment. The problem with this approach is that the resulting curve would have "corner points" and thus would not be smooth as desired. It would, however, be continuous. This curve is called a **linear spline** (or first degree spline). The linear spline for the 17 given data points is shown below.

❏ **6.3.** Indicate the result of **Enter**ing **linearfit[tbl,1,17]**

This is much better than the interpolating polynomial, isn't it? But it is not clear that it has corner points. Let's "zoom in" on the "hump" and get a closer look. Looking at the data, let's plot it between the data points **5** and **9** by **Entering linearfit[tbl,5,9].** Now we can clearly see that there are indeed corner points on the linear spline.

In order to connect successive pairs of points with "nice" curve segments to eliminate the corner points, we can consider using higher degree polynomials. If we use second degree polynomials, we obtain a **quadratic spline.** If we use third degree polynomials, we obtain a **cubic spline.** By connecting pairs of successive points with pieces of polynomials, the resulting interpolating function (spline) is continuous. Additional conditions will give the desired smoothness. Splines are very important in design work.

Let's examine the process of piecing parts of polynomials together to obtain a continuous function.

❑ **6.4.** If you used fourth degree polynomials, what kind of spline would you obtain?

EXAMPLE 7. Define $f(x) = \begin{cases} cx^2 - 3x + 5, x \geq 5 \\ x^2 - 5x + 7, x < 5 \end{cases}$

❑ **7.1.** For c=1, is f(x) continuous at x=5? (Justify your answer.)

❑ **7.2.** If not, is there a value of c that makes f continuous on [1,8]? Make sure that you can show on paper that your value of c is "as advertised". Is the resulting curve "smooth"? Why or why not? Be sure to show the graph. (HINT. See Example 2.)

❑ **7.3.** If there is a value of c as specified above, what kind of spline results? What is the set of data points?

Secant Lines and Tangent Lines

NAME(S):   INSTRUCTOR:
CLASS: DATE:

▣ INTRODUCTION

If h is a non-zero number and c and c+h are both numbers in the domain of y=f(x), then the slope of the line passing through the point (c,f(c)) and (c+h,f(c+h)) is given by

$$\text{slope}_{\text{secant line}} = \frac{f(c+h) - f(c)}{(c+h) - c} = \frac{f(c+h) - f(c)}{h}.$$

By definition, the slope of the tangent line to the curve y=f(x) exists at the point x=c only if the limit of the slopes of the secant lines between x=c and x=c+h converge to a fixed number:

$$\lim_{h \to 0} \text{slope}_{\text{secant line}} = \lim_{h \to 0} \frac{f(c+h) - f(c)}{h} = \text{slope}_{\text{tan line}}.$$

The purpose of this lab is to explore why this is the correct definition and compute slopes of lines tangent to certain functions.

▣ NEW *MATHEMATICA* COMMANDS AND PROCEDURES

The new *Mathematica* commands used in this notebook are:

1. **plotptdiscontinuity[f[x],{x,a,b},c]** graphs the function **f[x]** with a single discontinuity at x=c on the interval **[a,b]**.

2. **secantline[f[x],{x,a,b},c]** draws the graph of f on the interval [a,b], a dashed graph of the line tangent to the graph of f at the point (c,f(c)) and a sequence of secant lines about x=c. **secantline[f[x],{x,a,b},c,tan->no]** draws the graph of f on the interval [a,b] and a sequence of secant lines about x=c. This version should be used when the curve f **does not** have a tangent line at (c,f(c)).

These are not standard *Mathematica* commands. We have created these commands to use in this lab. Your instructor will make them available to you for this lab.

Animate graphics procedure. To Animate a Selection of Graphics Cells, **Select** all graphics cells to be animated, go to the **Graph** menu, and choose **Animate Selected Graphics**.

23

■ FUNCTIONS FOR WHICH THE TANGENT LINE EXISTS

> **EXAMPLE 1.** Define $f(x) = x^3 - 6x^2 + 11x - 6$ by executing the following commands:

Clear[f]
f[x_]:= x^3 - 6 x^2 + 11 x - 6

❑ **1.1.** Plot f(x) for $-3 \leq x \leq 6$.

❑ **1.2.** Execute command **secantline[f[x],{x,0,4},.9]** and then animate the set of pictures obtained. Keep your eye on the secant line. (That is the one that is changing).

❑ **1.3.** Re-execute the above command using: (a) 1 instead of .9; and (b) 1.1 instead of .9.

■ FUNCTIONS FOR WHICH THE TANGENT LINE DOES NOT EXIST

If a function is not continuous for a certain value of x, why doesn't the graph of the function have a tangent line at the point determined by that value of x? Let's look at some examples with this characteristic.

> **EXAMPLE 2.** (Jump Discontinuity) Define $f(x) = \begin{cases} x^2, & \text{if } x \leq 1 \\ 2x, & \text{if } x > 1 \end{cases}$ by executing the following commands.

Clear[f]
f[x_]:=x^2 /; x<=1
f[x_]:=2x /; x>1

NOTE. An alternative to defining f as above is **f[x_]:=If[x<=1, x^2, 2 x]**

❑ **2.1.** Plot f(x) on [-1.5,3] by **Entering plotptdiscontinuity[f[x],{x,-1.5,3},1]**
NOTE. There may be a discontinuity at x=1, thus we use **plotptdiscontinuity**.

Then graph some secant lines by **Entering secantline[f[x],{x,0,2},1,tan->no]**

❑ LAB # 4 — Secant/Tangent Lines ❑

❑ **2.2.** Use the graph to determine the limit of the slopes of the secant lines to the right of x=1.

Observe that this is the same as computing $\displaystyle \lim_{h \to 0^+} \frac{f(1+h) - f(1)}{h}$.

❑ **2.3.** Use the graph to determine the limit of the slopes of the secant lines to the left of x=1.

Observe that this is the same as computing $\displaystyle \lim_{h \to 0^-} \frac{f(1+h) - f(1)}{h}$.

❑ **2.4.** Why does the tangent line not exist at x=1?

EXAMPLE 3. (Point Discontinuity) Define $f(x) = \begin{cases} 2, & \text{if } x < 1 \\ 3, & \text{if } x = 1 \\ x, & \text{if } x > 1 \end{cases}$ by executing the following

commands .

Clear[f]
f[x_]:= 2 /; x<1
f[x_]:=3 /; x==1
f[x_]:= x /; x>1

❑ **3.1.** Using **plotptdiscontinuity** plot f on [-1,2].

Then graph some secant lines by **Entering secantline[f[x],{x,0,2},1,tan->no]**

❑ **3.2.** Use the graph to determine the limit of the slopes of the secant lines to the right of x=1.
HINT. See 2.2 above.

❑ **3.3.** Use the graph to determine the limit of the slopes of the secant lines to the left of x=1.
HINT. See 2.3 above.

25

❏ **3.4.** Why does the tangent line not exist at x=1?

❏ **3.5.** Does $\displaystyle\lim_{h\to 0}\frac{f(1+h)-f(1)}{h}$ exist?

If the limit does exist, what is the limiting value? If the limit does not exist, why not?

EXAMPLE 4. (Corner Point) $f(x)=\begin{cases} x^2, & \text{if } x\le 0 \\ 2x, & \text{if } x>0 \end{cases}$.

❏ **4.1.** After defining f, plot f(x) on [-1,1].

Then graph some secant lines by using **secantline**.

❏ **4.2.** Use the graph to determine the limit of the slopes of the secant lines to the right of x=0.

Observe that this is the same as computing $\displaystyle\lim_{h\to 0^+}\frac{f(0+h)-f(0)}{h}$.

❏ **4.3.** Use the graph to determine the limit of the slopes of the secant lines to the left of x=0.

Observe that this is the same as computing $\displaystyle\lim_{h\to 0^-}\frac{f(0+h)-f(0)}{h}$.

❏ **4.4.** Why does the tangent line not exist at x=1?

❏ **4.5.** Does $\displaystyle\lim_{h\to 0}\frac{f(0+h)-f(0)}{h}$ exist? If the limit does exist, what is the limiting value?
If the limit does not exist, why not?

Slopes, Tangent Lines
and
Derivatives

NAME(S):   INSTRUCTOR:
CLASS: DATE:

■ INTRODUCTION

Given a function f, then the function f' defined by

$$f'(x) = \lim_{h \to 0} \frac{f(x+h) - f(x)}{h},$$ provided this limit exists, is called the **derivative of f**.

A function f is **differentiable at a number w** means f'(w) exists; a function is **differentiable on an open interval (a,b)** (a may be -∞, b may be +∞) means that f'(w) exists for every number w in the interval (a,b).

If f is differentiable at w, then one way of interpreting f'(w) is that f'(w) is the slope of the line tangent to the graph of f at the point (w,f(w)).

The purpose of this lab is to compute derivatives of some functions and equations of some tangent lines. Applications of derivatives are covered in later labs.

■ NEW *MATHEMATICA* COMMANDS

The new *Mathematica* commands used in this lab are:

1. **f'[x]** computes the derivative (with respect to x) of the function **f** as long as **f** is a function of a single variable.

2. **D[f[x],x]** computes the derivative (with respect to x) of the function **f**.

■ COMPUTING DERIVATIVES AND SLOPES OF TANGENT LINES

As indicated above, *Mathematica* can compute the derivative of a function of a single variable in a variety of ways: if **f[x]** is a function of a single variable, the commands **f'[x]** and **D[f[x],x]** both

27

compute the derivative of f with respect to x.

EXAMPLE 1. Let $p(x) = 2 + x^2 - 3x^3 + 2x^4$.

First, carefully define p(x).

❏ **1.1.** Compute both **p'[x]** and **D[p[x],x]** by **Enter**ing the following commands. Compare your results.

p'[x]

D[p[x],x]

❏ **1.2.** Graph both p and p' on the interval [-.5,1] (p in black; p' in gray) by **Enter**ing:

Plot[{p[x],p'[x]},{x,-.5,1},PlotStyle->{GrayLevel[0],GrayLevel[.3]}]

❏ **1.3.** Use **NRoots** to approximate the values of x for which p'(x) has value 0.

❏ **1.4.** Use 1.3 to approximate the points (a,p(a)) for which the line tangent to the graph of p at the point (a,p(a)) is horizontal.

EXAMPLE 2. If $r(x) = \dfrac{-3 + 2x - 10x^2}{-6 + 10x}$, then we can compute the derivative of r(x) and write the results as a fraction with a single denominator.

First, define r(x):

Clear[r]
r[x_]=(-3+2x-10x^2)/(-6+10x)

28

❏ **2.1.** Compute r'(x) by **Entering:**

Together[r'[x]]

and by **Entering Together[D[r[x],x]]**

❏ **2.2.** For what values of x is the line tangent to the graph of r at the point (x,r(x)) horizontal?

❏ **2.3.** What is the domain of r' ?

EXAMPLE 3. Let $r(x) = \dfrac{-3 + 2x - 10x^2}{-6 + 10x}$. Find an equation of the line tangent to the graph of r at the point (x,r(x)) when $x = \dfrac{6}{5}$.

Carefully define r.

$r'\left(\dfrac{6}{5}\right)$ is the slope of the line tangent to the graph of r at the point $\left(\dfrac{6}{5}, r\left(\dfrac{6}{5}\right)\right)$.

❏ **3.1.** Compute both $r'\left(\dfrac{6}{5}\right)$ and $r\left(\dfrac{6}{5}\right)$ by **Entering r'[6 / 5]** and **r[6 / 5]**

r'[6/5]=

r[6/5]=

Therefore, an equation of the line tangent to the graph of r at the point $\left(\dfrac{6}{5}, \dfrac{-5}{2}\right)$ is

$y = \dfrac{1}{2}\left(x - \dfrac{6}{5}\right) - \dfrac{5}{2}$.

We can verify that y is indeed the desired line by defining y:

Clear[y]
y[x_]=1/2(x-6/5)-5/2

29

and then graphing both simultaneously on the interval [1,1.5] (r in black; y in gray):

Plot[{r[x],y[x]}, {x,1,1.5}, PlotStyle->{GrayLevel[0],GrayLevel[.3]}]

❑ **3.2.** Find and graph an equation of the line tangent to the graph of r at the point $(x, r(x))$ when $x = \dfrac{3}{4}$.

❑ **3.3.** For what values of x does the line tangent to the graph of r at the point $(x, r(x))$ have slope 1.   Graph r and all lines tangent to the graph of r with slope 1.

If the derivative of the first derivative of f exists, it is called the **second derivative of f** and can be computed with either **f''[x]** or **D[f[x],{x,2}]**

EXAMPLE 4. If $g(x) = 5\sin(6x) + 9\cos(14x)$, then we can compute the second derivative of g(x) after carefully defining **g** by **Entering:**

g''[x]

or by **Entering:**

D[g[x],{x,2}]

❑ **4.1.** Graph g(x) and g''(x), on separate axes, on the interval $\left[0, \dfrac{\pi}{2}\right]$.

❑ **4.2.** Use **FindRoot** to approximate the solutions of the equation g''(x) = 0 on the interval $\left[0, \dfrac{\pi}{2}\right]$.

30

Composition of Functions and The Chain Rule

NAME(S):   INSTRUCTOR:
CLASS: DATE:

▣ INTRODUCTION

One of the more powerful ways of building up complex functions from simpler functions is via composition. What is meant by the composition of the functions f(x) and g(x)? Technically, we define the composition of the two functions f and g, denoted by the symbol fog, pointwise as follows:

fog(x) = f(g(x))

i.e. f is evaluated at the point g(x). From this viewpoint, we can see that the composition arises naturally in many applications since we can view the composition of f and g as determining how the function f affects the output of the function g. For example, it is known that the profit P from producing x number of items of a commodity is a function of the price, p, of each item. (i.e. P=f(p)) But, the price of a particular item is frequently a function of the number of items produced. (i.e. p=g(x)) Thus, indirectly, the profit is a function of the number of items produced since P=f(p)=f(g(x)). (i.e. P=fog(x))

Since we can form interesting types of functions by composing two or more simple functions, it is natural to ask if the derivative of the composition of one or more functions can be defined in terms of the derivatives of the simple functions that make up the composition. The chain rule provides the answer to this question, namely, the derivative of the function formed from the composition of two or more functions is just the product of the derivatives of each of the functions that make up the composition. Of course, we have to be careful to evaluate the derivatives of each simple function at the correct point.

▣ NEW *MATHEMATICA* COMMANDS

To repeatedly apply a function f to a point x=a, we can use the **Nest** command as shown in the following example.

f[x_]:=Sin[x]
Nest[f,x,3]   (* Returns the value of f[f[f[x]]] *)

This is equivalent to composing f(x) with itself three times.

■ COMPOSITION OF FUNCTIONS

Some examples which follow show what types of functions can be constructed by forming the composition of simpler functions.

EXAMPLE 1. $f(x) = x^2$ and $g(x) = \sin(x)$.

❏ **1.1.** What is the domain of fog? of gof?

❏ **1.2.** Describe the shape of the curves fog(x) and gof(x). Note the algebraic difference between fog and gof.
SUGGESTION. Plot each curve on the same graph over the interval [-6,6]. Use the Plot option **PlotStyle->{GrayLevel[0],GrayLevel[.4]}** to distinguish each curve.

❏ **1.3.** What is the range of fog? of gof?

EXAMPLE 2. Composing sin(x) with itself several times.

What happens if you repeatedly apply, say n times, the sine function to the point x? That is, what does the graph of the function obtained by composing sin(x) with itself n times look like?

❏ **2.1.** Describe the difference between the curves $f(x) = \sin(x)$ and $g(x) = \sin(\sin(\sin(x)))$.
Note: The function g(x) is obtained from f(x) by composing f(x) with itself 3 times.
HINT: Define g using the **Nest** command, that is, define g by typing **g[x_]:=Nest[Sin,x,3]**

❏ **2.2.** Define h(x) by composing f(x) with itself 10 times. Plot f and h on the same graph. Describe the difference between the curves f(x) and g(x).

❏ **2.3.** Define k(x) by composing f(x) with itself 50 times. Plot f and k on the same graph. Describe the difference between the curves f(x) and k(x).

> **EXAMPLE 3.** Let $f(x) = \sqrt{x}$ and $g(x) = \sin(x)$.

❏ **3.1.** What is the domain of fog? of gof?

❏ **3.2.** Describe the shapes of the curves (fog)(x) and (gof)(x). Note the algebraic difference between fog and gof. (HINT. Plot the functions over the intervals [0,50] and [0,70].)

❏ **3.3.** What is the range of fog? of gof?

◾ USING *MATHEMATICA* TO FIND THE DERIVATIVE OF THE COMPOSITION OF FUNCTIONS

> **EXAMPLE 4.** Problems where both f(x) and g(x) are known.

Let $f(x) = \sin(x)$ and $g(x) = x^3 + 3x$

❏ **4.1.** Find the derivative of fog[x] by **Entering D[f[g[x]],x]**

Sometimes you might want to label this result so that you can refer to it later. You can do this by **Entering dfog = D[f[g[x]],x]** and then to evaluate this result at x= a, **Enter dfog /. x->a**

❏ **4.2.** Find d(fog(x)) at x=2.

❏ **4.3.** Find d(gof(x)) at x = 2.

EXAMPLE 5. Problems where f(x) and g(x) are not known but their derivatives are known.

❏ **5.1.** A fire has started in a dry open field and spreads in the form of a circle. The radius of the circle increases at the rate of 6ft./min. Find the rate at which the fire area is increasing when the radius is 100 feet.

❏ **5. 2.** A paper cup containing water has the shape of a right circular cone of altitude 6 inches and radius 2 inches . If water is leaking out of the cup at the rate of 3 in.^3 /hr, at what rate is the water level decreasing when the depth of water is 4 inches?

EXAMPLE 6. Dependence of the speed of a pendulum on the length of the pendulum.

Suppose that the angle θ at time t(in seconds) of a pendulum of length **lth** is given by the equation

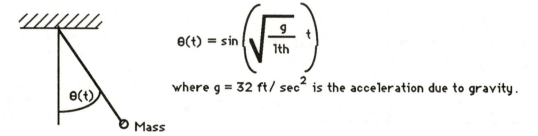

$$\theta(t) = \sin\left(\sqrt{\frac{g}{lth}}\; t\right)$$

where $g = 32 \text{ ft/sec}^2$ is the acceleration due to gravity.

❏ **6.1.** Graph both θ(t) and θ'(t) for lth = 1, 2, and 4. Describe how the graphs of θ(t) and θ'(t) change as the length lth changes.

❏ **6.2.** If lth doubles, does the pendulum swing faster or slower? Give an argument to support your answer. (HINT. At what point is the speed of the pendulum the fastest?)

Related Rates
and
Implicit Differentiation

NAME(S):   INSTRUCTOR:
CLASS: DATE:

■ INTRODUCTION

One topic to be considered in this lab is **related rates**. In solving a related rate problem, one finds the rate of change of some quantity by relating this quantity to other quantities whose rates of change are known.

The other main topic to be studied is **implicit differentiation**. Quite often one wants to graph a relation in x and y that is not a function of x. It may be impractical to break-up this relation into functions of x, but suppose that we can represent it as $f(x,y) = 0$ and graph the equation in that form. Additionally, a typical question concerns finding tangents to the resulting curve. In implicit differentiation, we assume that $y = y(x)$ at points under consideration. Then we take the derivative of both sides of the equation $f(x,y) = 0$ with respect to x. One then solves the resulting equation for $y'(x)$. This expression is generally an expression in x and y. With a point on the curve, one can then find $y'(x)$ at that point. Thus, knowing a point of contact and the slope of the tangent line (the slope is $y'(x)$), one can find the equation of that tangent line using the point-slope formula for a line.

■ NEW *MATHEMATICA* COMMANDS

One function that is very useful has the form **expression/.variable->constant**. In **expression**, every occurrence of **variable** is replaced by **constant**. Thus, we say **expression** is evaluated by the rule **variable->constant**. Other rules may be adjoined as follows: expression/.var1->con1/.var2->con2, etc. In this case, the rule var1->con1 is applied first, rule var2->con2 is applied next, etc. For example, **x^2+2 x y+y^2/.x->2** yields 4+4y+y^2 while **x^2+2 x y+y^2/.x->2/.y->-1** yields 1.

If rule = {x->a}, then **expression/.rule** is the same as expression/.x->a. For example, if rule = {x->2}, then **x^2+2 x y+y^2/.rule** yields 4+4y+y^2.

Let f be a function of x and y, then z=f(x,y) is a 3-dimensional surface. To graph f(x,y)=0, we use ContourPlot. For example, the command

ContourPlot[f[x,y],{x,a,b},{y,c,d},PlotRange->{0,0},PlotPoints->n]

graphs f(x,y)=0 for a≤x≤b and c≤y≤d. **PlotRange->{0,0}** guarantees f(x,y)=0 while **PlotPoints->n** indicates that a minimum of n points will be used to sample the function. The default is n=15. In some cases, you will need to have n>15 to get the desired accuracy. You should expect the graphing to take somewhat longer if you choose n much bigger than 25.

If S is a set, then S[[i]] indicates the i-th element of S.

Chop[expression] replaces all real approximate numbers in **expression** having absolute value less than 10^{-10} with the value 0.

▣ PLOTTING f(x,y)=0

EXAMPLE 1 . The graph of $x^{\frac{2}{3}} + y^{\frac{2}{3}} = 1$ is called a four - cusped hypocloid.

❑ **1.1.** Plot the graph for $-1 \leq x \leq 1$, $-1 \leq y \leq 1$. Use the default for PlotPoints.

HINT. First put the equation in the form f(x, y) = 0. (i.e. $f(x, y) = x^{\frac{2}{3}} + y^{\frac{2}{3}} - 1$.)

▣ RELATED RATES

EXAMPLE 2. Suppose $v(t) = \pi \, (r(t))^2 \, h(t)$ and, for a certain value of t, h'(t) = 2, r'(t) = 6, r(t) = 3, and h(t) = 5.

First define v(t) by **Clearing v, r,** and **h,** and then **Entering v[t_]:=Pi r[t]^2 h[t]**

❑ **2.1.** Find v'(t), for arbitrary t, by **Entering vprime=D[v[t],t]**

v'(t)=

❑ **2.2.** Now, find v'(t) at that certain value of t by **Entering**
result=vprime/.h'[t]->2/.r'[t]->6/.r[t]->3/.h[t]->5

result=

❑ **2.3.** Now, convert the result to decimal form by **Entering result//N**

decimal form of result =

❑ **2. 4.** Define $w(t) = d(v(t))$, where $d(t) = t^2 + 3t - 1$. Find $w'(t)$ in general and at that certain value of t.

> **EXAMPLE 3.** In chemistry, there is a process in which there is no gain or loss of heat. During this process, called an adiabatic process, the pressure p and volume v of certain gases such as oxygen or hydrogen in a container are related by the formula $p v^{1.4} = $ constant. At a certain time, the volume of oxygen in a closed container is 12 m^3 and the pressure is .93 kg / m^3. Suppose the pressure is increasing at a rate of .31275 kg / m^2 / sec. What is the rate of change of the volume?

Since pressure and volume are functions of time, p=p(t) and v=v(t). We have information about v(t), p(t) and p'(t) at a certain time t. The unknown is v'(t) at that time.

❑ **3.1.** First, **Clear** p and v, and then find the derivative of the left-hand-side of the equation by **Entering deriv=D[p[t] v[t]^1.4,t]**

deriv=

❑ **3.2.** Since the right-hand-side of the equation is constant, its derivative is 0. Thus, we want to solve **deriv=0** for v'(t). To do so, **Enter yprime=Solve[deriv==0,v'[t]]**

yprime=

❑ **3.3.** At certain time t, p(t)=.93, v(t)=12, and p'(t)=.31275. State the command to find v'(t) at that certain time and state the resulting answer.

command=

v'(certain time)=

❑ **3.4.** State the complete answer to the problem giving the units.

complete answer=

❑ **3.5.** In this example, instead of the **pressure increasing**, suppose the **volume is decreasing** at a rate of .13725 kg/m. What is the rate of change of the pressure? State your steps and state the complete answer.

> **EXAMPLE 4.** The thin lens equation in physics is $\dfrac{1}{od} + \dfrac{1}{id} = \dfrac{1}{fl}$, where od is the object distance from the lens, id is the image distance from the lens, and fl is the focal length of the lens. Suppose that a certain lens has a focal length of 7 cm and that an object is moving toward the lens at the rate of 2.5 cm / sec. How fast is the image distance changing at the instant when the object is 12 cm from the lens?

Note that fl is constant while od=od(t) and id=id(t). (i.e. Both od and id are functions of time.) We are asked to find id'(t) at a certain instant, thus we need to find an expression for it. Since fl is constant, the derivative of the right-hand side of the equation is 0. Thus, we take the derivative of the left-hand side of the lens equations with respect to t, set it equal to 0 and solve for id'(t).

❏ **4.1.** Find id'(t).

id'(t)=

❏ **4.2.** Find id'(t) at that certain instant and determine whether the image is moving toward the lens or away from it.

id'(certain instant)=

▪ IMPLICIT DIFFERENTIATION

Suppose we have a relation of the form f(x,y)=g(x,y) and we want to find y' by implicit differentiation. (We assume y=y(x).) Using *Mathematica*, we take the derivative of each side (i.e.D[f[x,y[x]],x] and D[g[x,y[x]],x]), equate them, and solve for y'[x] by using the Solve command. Putting these two steps together, we would get Solve[D[f[x,y[x]],x]==D[g[x,y[x]],x],y'[x]]. Let's see if we can accomplish this with one D operator. Since *Mathematica* cannot perform D[exp1==exp2, y'[x]], we put f(x,y)=g(x,y) in the form f(x,y)-g(x,y)=0. Then we can get the desired result as follows:
Solve[D[f[x,y[x]]-g[x,y[x]],x]==0,y'[x]]

EXAMPLE 5.   If sin(y)+sin(x)=1, find y'.

Let's put the equation in the form f(x,y)=0. Then we have sin(y)+sin(x)-1=0. To define f(x,y),
Enter the commands   **Clear[f]**
                              f[x_,y_]:=Sin[y[x]]+Sin[x]-1

❏ **5.1.** To find y', **Enter**   **yprime=Solve[D[f[x,y],x]==0,y'[x]]**

yprime=

❏ **5.2.** What is yprime? Notice that it is a set having one element--a rule of the form {y'[x]->**expression in x and y**}. Since yprime has only one element, that element is yprime[[1]]. Now let's evaluate y'[x] according to that rule. **Enter slope=y'[x]/.yprime[[1]]**

slope=

Another way to do the above computation is to use the **Dt** operator. Using this operator, one does not have to formally declare that y=y(x). **Dt** stands for total derivative and you will not study this topic until much later, however, we will demonstrate how it is used to find derivatives implicitly. In the following, Dt[y,x] will be y'. Enter each of the following and compare the result to the corresponding result obtained above.

❏ **5.3. Enter yprime=Solve[Dt[Sin[y]+Sin[x]-1,x]==0,Dt[y,x]]**

yprime=

❏ **5.4. Enter slope=Dt[y,x]/.yprime[[1]]**

slope=

> **EXAMPLE 6.** $\sqrt{1 + \sin^3\left(xy^2\right)} = y$

❏ **6.1.** Find y' by implicit differentiation.

y'=

> **EXAMPLE 7.** The graph of the equation $4(x^2 + y^2)^2 = 41(x^2 - y^2)$ is called a lemniscate .

We want to graph the equation and then determine how to find tangent lines to that graph. First, put the equation in the form f(x,y)=0 and then clear and define f. i.e. **Enter**
Clear[f]
f[x_,y_]:=4 (x^2 + y^2) ^2 - 41 (x^2 - y^2)

❏ **7.1.** To obtain the desired graph, we plot it (and save it) as follows. (NOTE: Recall that **PlotPoints->15** is the default. We are using **PlotPoints->25** to get a better graph. If you change 25 to 50, you will get a more accurate graph but it will take much longer to get it.) **Enter** the following and sketch the resulting graph.
graph=ContourPlot[f[x,y],{x,-4,4},{y,-4,4},PlotRange->{0,0},PlotPoints->25]

If you like, you can later show the graph again by **Entering Show[graph]**

❏ **7.2.** In order to find tangent lines to the graph, we assume that y = y(x) on some interval about the x-value being considered. Thus, we need to find y'. As mentioned before, it is not practical to solve the equation for y explicitly and then take its derivative. So, we work directly with the equation and assuming y=y(x), we take the derivative implicitly. First, we will find y'. Then, we will evaluate it at some point on the graph of f(x,y)=0. Using **5.1** as a guide, find y'.

yprime=

❏ **7.3.** Using **5.2** as a guide, find **slope**.

slope=

❏ **7.4.** Since slope is a function of x and y, we must have a point on the graph before we can find the slope of the tangent line to the graph. (If it has a tangent line there!) Let's find what point(s) on the graph correspond to an x-value of 1.5. **Enter** the following commands and state your result.

xvalue=1.5
expression=f[xvalue,y]

❏ **7.5.** In order to find the corresponding y-value(s), we solve the equation **expression = 0** for y as follows. **Enter yvalue=Solve[expression==0,y]** and state the result.

yvalue=

Note that yvalue is a list of rules of the form **{y ->value}**. You can see that two of the rules correspond to real values of y. (Of course, we are not interested in the non-real values of y.) Since we have two real values of y, there are two points on the graph corresponding to xvalue =1.5.

❏ **7.6. Enter yvalue=y/.yvalue[[1]]** and state the result.

yvalue=

❏ **7.7.** The expression y was evaluated according to the first rule in yvalue. To verify that (xvalue,yvalue) is on the graph of f(x,y) = 0, let's substitute these values into the expression f(x,y). **Enter f[xvalue,yvalue]** and state the result. You probably won't get **exactly** 0. Why not?

f[xvalue,yvalue]=

❏ **7.8. Enter Chop[f[xvalue,yvalue]]** and explain the result.

❏ **7.9.** To find the slope m of the tangent line to the graph of f(x,y) = 0 at (xvalue,yvalue), we just evaluate the expression **slope** at that point. **Enter m=slope/.y[x]->yvalue/.x->xvalue** and state the result.

m=

❏ **7.10.** The order of this evaluation is important. The above order means that y[x]->yvalue is done first. Let **m1=slope/.x->xvalue/.y[x]->yvalue** and state the value of m1. Does m1 = m?

m1=

❏ **7.11.** Now, we want to find the line passing through (xvalue, yvalue) with slope m. Using the point-slope formula for a line, the equation is y=m(x-xvalue)+yvalue. **Enter result=m (x-xvalue)+yvalue** to obtain the equation.

result=

❏ **7.12.** Now **Simplify[result]** and state the equation of the tangent line and the point on the lemniscate through which it passes.

❏ **7.13.** Find the equation of the tangent line corresponding to the other real y-value.

Graphing Functions Using Properties of the Derivative

NAME(S):   INSTRUCTOR:

CLASS: DATE:

▣ INTRODUCTION

If f is a differentiable function, then properties of the first and second derivative yield information about the behavior and graph of f. For example, the values of x for which f'(x) is positive correspond to those values of x for which f(x) is (strictly)increasing; the values of x for which f'(x) is negative correspond to those values of x for which f(x) is (strictly)decreasing; the values of x for which f"(x) is positive correspond to those values of x for which f(x) is concave up; and the values of x for which f"(x) is negative correspond to those values of x for which f(x) is concave down.

The purpose of this lab is to use the information the first and second derivatives of a function yield to obtain information about the original function and its graph.

▣ NEW *MATHEMATICA* COMMANDS AND TERMINOLOGY

The new *Mathematica* commands used in this lab are:

1. Numerator[f[x]] returns the numerator of **f[x]**.

2. Denominator[f[x]] returns the denominator of **f[x]**.

3. z is a **critical number** of f means that either f'(z)=0 or f'(z) does not exist.

4. f is **monotone increasing** means that f(x)<f(y) whenever x<y (visually, this means that the graph of f goes up as you go from left to right); f is **monotone decreasing** means that f(x)>f(y) whenever x<y (visually, this means that the graph of f goes down as you go from left to right); and f is **monotone** means that f is either monotone increasing or monotone decreasing.

To find the values of x for which f is increasing or decreasing:
  1. Find the values of x for which **f'(x) is positive**.
     (this is the same as the values of x for which f is increasing.)
  2. Find the values of x for which **f'(x) is negative**.
     (this is the same as the values of x for which f is decreasing.)

To find the values of x for which f is concave up or concave down:
1. Find the values of x for which **f''(x) is positive**.
   (this is the same as the values of x for which f is concave up.)
2. Find the values of x for which **f''(x) is negative**.
   (this is the same as the values of x for which f is concave down.)

▣ EXAMPLES

EXAMPLE 1. If $f(x) = \dfrac{3x^5 - 5x^3}{6}$, find the values of x for which f is

(i) increasing; (ii) decreasing; (iii) concave up; and (iv) concave down.

How can we approach this problem? Let's see how we can use information about the first and second derivatives of f to answer these questions. First, clear f and define f by **Entering** the following commands:

Clear[f]

f[x_]:=(3x^5-5x^3)/6

❏ **1.1.** To obtain information about where f is increasing(decreasing), graph f(x) and f'(x) simultaneously on [-3,3]. In addition to giving the graphs, state the command you use.

❏ **1.2.** The values of x at which f'(x) changes from positive-to-negative or negative-to-positive are critical numbers. (Note that if x is a critical number, then either f'(x)=0 or f'(x) does not exist.) Enter **Solve[f'[x]==0]** to determine where f'(x)=0.

Since the change in behavior of f occurs within the interval [-1,1], replot f on the interval [-1.3,1.3] and answer the following questions.

❏ **1.3.** On what intervals is f increasing?

❏ **1.4.** On what intervals is f decreasing?

❏ **1.5.** What are the critical numbers?

❏ **1.6.** To obtain information about the concavity of f, graph f(x) and f"(x) simultaneously on [-1.3,1.3]. In addition to giving the graphs, state the command you use.

❏ **1.7.** The values of x at which f"(x) changes from positive-to-negative or negative-to-positive are the values of x for which either f"(x)=0 or f'(x) does not exist. **Enter** the following commands and indicate the results.

Solve[f''[x]==0]

Solve[f''[x]==0]//N

❏ **1.8.** On what interval(s) is f concave up?

❏ **1.9.** On what interval(s) is f concave down?

❏ **1.10.** What are the inflection points of f?

EXAMPLE 2. Discuss the behavior of $h(x) = 3\sqrt[3]{x^2}\,(3x - 7)^2$.

Begin by clearing all prior definitions of h and then carefully defining h.

❏ **2.1.** Graph h(x) and h'(x) on [-3,5] and indicate the result.

❏ **2.2.** On what interval(s) is h increasing ? decreasing ?

❏ **2.3.** What are the critical numbers?

❏ **2.4.** To obtain information about the concavity of h, graph h(x) and h"(x).
HINT. Don't forget to find the values of x for which h"(x) does not exist. (The command **Together[h''[x]]** may be useful.)

43

❑ **2.5.** On what interval(s) is h concave up? concave down?

❑ **2.6.** What are the inflection points ?

EXAMPLE 3.   Discuss the behavior of $r(x) = \tan\left(x^2 + 1\right)$.

Begin by clearing all prior definitions of r and then carefully defining r.

❑ **3.1.** Graph r(x) and r'(x) on [-5,5].

❑ **3.2.** On what intervals is r increasing? decreasing?

❑ **3.3.** What are the critical numbers?   HINT. Be careful; consider what the domain of r is.

❑ **3.4.** Where are the asymptotes located for the x values between -5 and 5?
HINT. Use the **Solve** command.

❑ **3.5.** To obtain information about the concavity of r, graph r(x) and r"(x).

❑ **3.6.** Identify the intervals between -5 and 5 on which r is concave up(down).

Applied Max/Min Problems (one variable)

NAME(S):   **INSTRUCTOR:**
CLASS: **DATE:**

■ INTRODUCTION

Having access to a tool such as calculus allows us to obtain answers to questions(problems) that seek to find the "best"answer. Frequently the "best" answer involves finding the maximum value, such as the maximum profit or the maximum level of harvest, or finding the smallest(minimum) answer such as the minimum cost of production or the minimum amount of pollution. Questions of this sort involve two separate issues. One issue involves determining the mathematical model that describes the situation being studied and the second issue involves using calculus tools to find the desired answer. In the problems below, we will consider some examples that involve setting up a mathematical model to find the "best" answer to a particular question.

■ NEW *MATHEMATICA* COMMANDS

There are no new *Mathematica* commands introduced in this lab.

■ MAX/MIN PROBLEMS

> **EXAMPLE 1. Maximizing Income.** A real estate company owns 180 apartments which are fully occupied when the rent is $300 per month. The company estimates that for each $10 increase in rent, five apartments will become unoccupied.

❏ **1.1.** What would the income be if the company charged $300 rent?

❏ **1.2.** What would the income be if the company charged $340 rent?

❏ **1.3.** What would the income be if the company charged $360 rent?

❑ **1.4.** What rent should be charged to obtain the largest gross income? As well as stating the answer, indicate the steps you took in order to obtain the solution.

EXAMPLE 2. Minimizing Surface Area. A cylindrical can with a volume of 218 cubic inches is to be made by cutting its top and bottom from metal squares and forming its curved surface by bending a rectangular sheet of metal into a circle.   What radius r and height h will minimize the amount of material required?

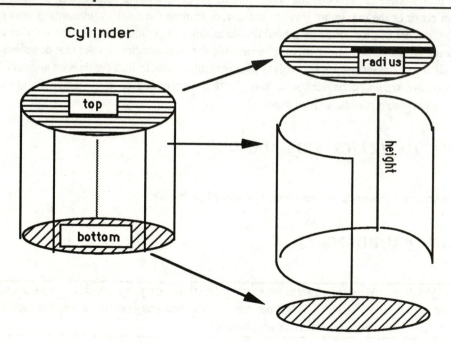

Cylinder

NOTE 1. This is a problem that asks you to find the minimum surface area subject to the side condition(constraint) that the volume is constant.

NOTE 2. The surface area can be broken up into three separate parts, namely,
SurfaceArea(can) = SurfaceArea(top) + SurfaceArea(side) + SurfaceArea(bottom)
= SurfaceArea(side) + 2 SurfaceArea(top).
The preceding figure shows this separation.

❑ **2.1.** What is the area of the side, expressed in terms of r and h?

❏ **2.2.** What is the area of the top, expressed in terms of r and h?

❏ **2.3.** What is the function, call it sa(r,h), that describes the total surface area of the can?
NOTE. This function can be expressed in terms of both r(adius) and h(eight).

❏ **2.4.** What is the value of r and h that will minimize the function sa?
HINT. Eliminate either h or r, and then use calculus to find the minimum value.

EXAMPLE 3. Ye Ole Pigeon Problem.

It is well known that homing pigeons avoid flying over large areas of water unless they are forced to do so. The reason for this behavior is not known at the present time. In our example we suppose that the pigeons prefer a detour around a lake since at daytime the air is falling over cool water, phenomena which increases the energy required for maintaining altitude over flight. In the figure below we assume that a pigeon is released from a boat at point B floating on the west side of a lake, whereas the loft point, point L, is located on the south-east bank. Assume that the energy required for flying one unit of length over the lake is e1 and the energy required for flying along the bank is c*e1 for some constant c>1. Suppose, instead of taking the shortest route from B to L, the pigeon makes a detour and heads to a certain point P on the southern bank and then eastward along the bank to L. For simplicity, we assume that the bank is straight in the east-west direction. The question arises: Where should the point P be chosen in order to minimize the energy required for the flight from B to L. What is the optimal angle BPL?

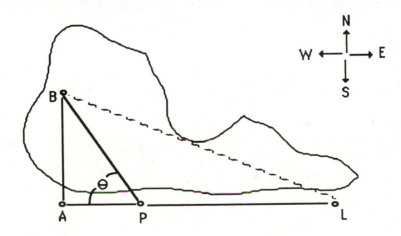

❑ **3.1.** What is the function, in terms of the angle $\theta = \angle APB$, that gives the total energy required to fly from B to P and then from P to L?

❑ **3.2.** Find the optimal angle APB that will minimize the energy required for flight from B to L.

EXAMPLE 4. A Shortest Path Problem

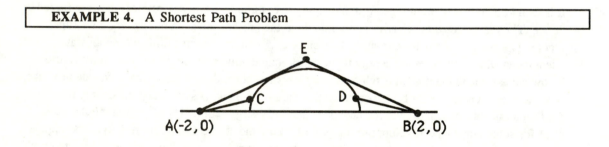

❑ **4.1.** Find the shortest path from $A = (-2,0)$ to $B = (2,0)$ avoiding the region D: $x^2 + y^2 \leq 1$.
(i.e. Walk past a circular lake without getting your feet wet.)

Check all straight segment paths such as path AEB and circular paths such as path ACDB. As well as stating the answer, indicate the steps you took in order to obtain the solution.

EXAMPLE 5. A billboard 20 feet tall is located on the side of a road with its lower edge 60 feet above the level of the center of a highway, as shown in the figure below. If the viewer stood in the middle of the road then this value could be used to determine the distance from the road that the sign should be placed.

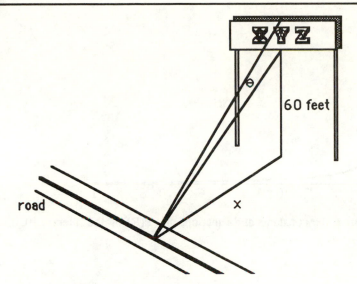

❑ **5.1.** How far from a point directly below the sign should a viewer stand to maximize the angle θ between the lines of sight of the top and bottom of the billboard? As usual, in addition to the answer, indicate the steps you take in obtaining the answer.

EXAMPLE 6. A device (battery or generator, for instance) that supplies electrical energy is called a source of **electromotive force** (abbreviated **emf**). Suppose a battery has emf F and an internal resistance r. If the external resistance of the circuit is R, then the current in the circuit is $I = \dfrac{F}{r + R}$ and the power delivered to the external resistance is $P = I^2R = \left(\dfrac{F}{r + R}\right)^2 R$. Thus, when F and r are constant, P is a function of R. (i.e. $P = P(R)$)

Suppose a car battery has F=12 and r=.0025.

❏ **6.1.** Plotting P(R) on the interval [0,1], we obtain the following. Does P achieve a maximum?

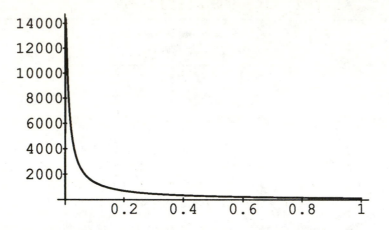

❏ **6.2.** After defining the constants and function, plot P(R) on the interval [0,.05].

❏ **6.3.** What property do you see in this graph that you did not see in the previous graph? Explain why. There is a lesson here for you to learn. What is it?

❏ **6.4.** Find the maximum value of P on [0,+∞) and the value of R where it is achieved. Indicate commands you use as well as the results you obtain.

Now, suppose F and r are positive constants but their actual values are not given to you.

❏ **6.5.** Find the maximum value of P on [0,+∞) and the value of R where it is achieved.

Antidifferentiation

NAME(S):   INSTRUCTOR:
CLASS: DATE:

▣ INTRODUCTION

The main topic of this lab is **antidifferentation**, which is the process of finding **antiderivatives**. Recall that F(x) is an antiderivative of f(x) if F'(x)=f(x). Note that if F'(x)=f(x), then (F(x) +c)'=F'(x). Thus, if F(x) is an antiderivative of f(x), then F(x)+c is also an antiderivative of f(x).

The family of all antiderivatives of $f(x)$ is called the **indefinite integral** of $f(x)$ and is denoted by $\int f(x)dx$. If F(x) is any antiderivative of $f(x)$, then we say $\int f(x)dx = F(x) + C,$ where C is any constant. C is called the constant of integration. Recall that $\int$ is a linear operator in the sense that $\int cf(x)dx = c\int f(x)dx$ and $\int (f(x) + g(x))dx = \int f(x)dx + \int g(x)dx.$

If you consider the family of antiderivatives of a function f(x), what is the **algebraic relationship** between any two of these antiderivatives? What is the **graphical relationship** between any two of these antiderivatives?

Recall that not all functions have antiderivatives in closed form (not even all continuous ones!). Since all Computer Algebra Systems(CAS) have limitations, there will be instances where a function has an antiderivative but a CAS can't find it. Thus, *Mathematica* will not always be able to find what you want. Additionally, *Mathematica* has "cooked up" some special functions that you may encounter during integration. For example, **Erf(x)** is a function which itself is defined as an integral. Thus, you need to be aware that even if you get an answer when attempting to find an indefinite integral, it may not be meaningful.

▣ NEW *MATHEMATICA* COMMANDS

The only new *Mathematica* command to be used in this lab is the **Integrate** command. The general form is **Integrate[f[var],var].** If f(var) is sufficiently simple, it returns the indefinite integral of f(var) (without the constant of integration). Thus, **Integrate[f[var],var]** actually returns the antiderivative of f(var) which has the constant term zero. The other commands used will be commands you have already encountered. However, since some of you may be using these labs for the first time, we will provide more assistance than usual with commands previously used. If you are using this lab manual now for the first time, you will probably want to go through LAB # 1 as it will be very useful to you. If you are using Version 1 of *Mathematica*, you will need to load IntegralTables before you **Enter** the Integrate command. To do so, **Enter** the command **<<IntegralTables.m**

EXAMPLE 1. $f(x) = 3x^2 + 6x + 2$

First, define f(x) by **Enter**ing the following commands.
Clear[f]
f[x_]:=3 x^2 + 6 x + 2

❏ **1.1.** Indicate the result of **Enter**ing the following commands:

?f

f[x]

f[1]

f[-1]

❏ **1.2.** Graph f(x) on [-2,.5], and save it as **graph1**, by **Enter**ing
graph1=Plot[f[x], {x,-2,.5}, PlotStyle->GrayLevel[0]]

❏ **1.3.** Find antif(x), the antiderivative of f(x), by **Enter**ing the following commands:
Clear[antif]
antif[x_]=Integrate[f[x],x]

Complete: $\int f(x)dx =$

❏ **1.4.** Now, plot antif(x) on [-2,.5], and save it as **graph2**, by **Enter**ing
graph2=Plot[antif[x], {x,-2,.5}, PlotStyle->GrayLevel[.4]]

❏ **1.5.** Now find the derivative of antif(x) by **Enter**ing **D[antif[x],x]**
Is the result as you expected? Why or why not?

❏ **1.6.** Does this graph have any particular relationship to the graph of f(x)? Perhaps it would be useful to see both graphs together. To accomplish that, **Enter Show[graph1,graph2]** Now, sketch the graphs and answer the question.

❏ **1.7.** Plot antif(x) and antif(x)+3 on [-2,.5] by **Entering**
Plot[{antif[x],antif[x]+3}, {x,-2,.5}, PlotStyle->{GrayLevel[0], GrayLevel[.4]}]

❏ **1.8.** What is the relationship between the two functions antif(x) and antif(x)+3:

Algebraically?

Graphically?

❏ **1.9.** Suppose that g(x) and h(x) are both antiderivatives of f(x). What is the relationship between g(x) and h(x):

Algebraically?

Graphically?

❏ **1.10.** Suppose that g(x) and h(x) are both antiderivatives of f(x). Find:

g'(x)

h'(x)

EXAMPLE 2.   g(y)=sin(y) cos(y)

❏ **2.1.** First define g by **Enter**ing:

Clear[g]

g[y_]:=Sin[y] Cos[y]

Now, plot g on [-π,π] by **Enter**ing **Plot[g[y],{y,-Pi,Pi}]**

❏ **2.2.** Indicate the results of **Enter**ing the following commands:

g[Pi/3]

N[g[Pi/4],10]

❏ **2.3.** Find the indefinite integral **antig(y)** of g(y).

$$\int g(y)dy \ = $$

❏ **2.4.** Now, find the derivative of **antig(y)** (with respect to y). Is the result as you expected? Why or why not?

EXAMPLE 3.   h(x) = x² sin(x)

First, carefully define h(x).

❏ **3.1.** Plot h(x) on [0,4π].

❏ **3.2.** Find $h\left(\dfrac{\pi}{2}\right)$ to 10 decimal places. (HINT. See **2.2.**)

$$h\left(\frac{\pi}{2}\right) =$$

❏ **3.3.** Find $\int h(x)dx$.

$$\int h(x)dx =$$

❏ **3.4.** Find $D_x\left[\int h(x)dx\right]$

$$D_x\left[\int h(x)dx\right] =$$

EXAMPLE 4. $f(y) = \dfrac{y^2}{\left(y^2+1\right)\left(y^2+3\right)}$

❏ **4.1.** Find $antif(y) = \int f(y)dy$.

$$\int \frac{y^2}{\left(y^2+1\right)\left(y^2+3\right)}dy =$$

❏ **4.2.** Find $D_y(antif(y))$.

$$D_y(antif(y)) =$$

❏ **4.3.** Indicate the result of **Entering Together[D[antif[y],y]]**. Does this differ from the result in **4.2** above? If so, how?

EXAMPLE 5. $g(x) = e^{-x^2}$

Carefully define g(x).

❏ **5.1.** Find the indefinite integral **antig(x)** of g(x).

$$\int e^{-x^2}dx =$$

❏ **5.2.** Find $D_x[antig(x)]$.

$$D_x[antig(x)] =$$

55

❑ **5.3.** Using your result from **5.1** above, find **Erf(x)**.

Erf(x)=

EXAMPLE 6. Since F(x) is an antiderivative of f(x) if and only if F'(x)=f(x), every formula for a derivative generates an integral formula. (This gives you a way to check your answer to an integral.)

The formula $D_x[\tan(x)] = \sec^2(x)$ gives rise to the integral formula $\int \sec^2(x)dx = \tan(x) + C$

and the formula $D_x[x^{\frac{1}{2}}] = \frac{1}{2}x^{-\frac{1}{2}}$ gives rise to the integral formula $\int \frac{1}{2}x^{-\frac{1}{2}}dx = x^{\frac{1}{2}} + C.$

❑ **6.1.** Take the derivative of sin(x) - x cos(x) and state the integral formula it generates.

EXAMPLE 7. **(Boundary Value Problem)** Find y=y(x) that satisfies:
$$y'=x - x\sin(x) + 1 \text{ and } y(\pi)=1.5$$

❑ **7.1.** To find y(x), we simply integrate y'. Define yp(x) to be y' as indicated above.

❑ **7.2.** Now **Enter Integrate[yp[x],x]** to obtain y(x). (Don't forget to add the constant C of integration.)

y(x)=

❑ **7.3.** Now impose the condition y(π)=1.5 and **Solve** for C. Substituting this value for C, the solution to the **BVP** is

y(x)=

EXAMPLE 8. (Another **BVP**). Solve for y:
$$y''=x\cos(x) + 2x - 3; \ y'(3)=1; \text{ and } y(3)=2.$$

❑ **8.1.** Using **7.** above as a guide, find y(x).
HINT. First find y'(x) and then find y(x). (i.e. It will require two integrations.)

EXAMPLE 9. At each point (x,y) on a curve, the slope of the tangent line is 3x-sec(x) tan(x). If the curve passes through the point (1,2), find the equation of the curve.

❑ **9.1.** Recall that if y=y(x) is the equation of the curve, then you only need to find y(x) that satisfies y'=3x - sec(x) tan(x) and y(1)=2. Now find y(x).

Riemann Sums

NAME(S):   INSTRUCTOR:

CLASS: DATE:

■ INTRODUCTION

Let f be a continuous function on an interval $[a,b]$. Given a positive integer N, partition $[a,b]$ into N equal subintervals:

$$[a,b] = [a, a + \frac{b-a}{N}] \cup [a + \frac{b-a}{N}, a + \frac{2(b-a)}{N}] \cup [a + \frac{2(b-a)}{N}, a + \frac{3(b-a)}{N}] \cup ..$$

$$.. \cup [a + \frac{(i-1)(b-a)}{N}, a + \frac{i(b-a)}{N}] \cup ... \cup [a + \frac{(N-1)(b-a)}{N}, b].$$

For convenience, call the ith subinterval, $[a + \frac{(i-1)(b-a)}{N}, a + \frac{i(b-a)}{N}]$, $[a,b]_i$.

Randomly pick a point x_i^* in each $[a,b]_i$. Let $RS(N; x_1^*, x_2^*, x_3^*, ..., x_N^*)$ denote

$\frac{b-a}{N} \sum_{i=1}^{N} f(x_i^*)$. The **integral of f on the interval [a,b]** is defined to be

$\underset{N \to \infty}{\text{Lim}} RS(N; x_1^*, x_2^*, x_3^*, ..., x_N^*)$, if this limit exists, and is denoted by $\int_a^b f(x)\,dx$.

Sums of the form $RS(N; x_1^*, x_2^*, x_3^*, ..., x_N^*) = \frac{b-a}{N} \sum_{i=1}^{N} f(x_i^*)$ are called **Riemann sums**.

Note that if $\int_a^b f(x)\,dx$ exists, then it is the limit of a sequence of Riemann sums. Thus, each

Riemann sum is an approximation of $\int_a^b f(x)\,dx$.

If the integral of f on the interval $[a,b]$ exists, the function is said to be **integrable on [a,b]**.

When f is a continuous function on [a,b], the integral of f on [a,b] ALWAYS exists!!

The purpose of this lab is to experiment with various Riemann sums so that we will better understand this very important concept.

■ NEW *MATHEMATICA* COMMANDS

The new *Mathematica* command used in this lab is **riemannsum**. The command
riemannsum[f[x],{x,a,b},n,pt,graph] computes a Riemann sum for **f** by dividing [**a,b**] into **n**
equal subintervals, picking **pt** inside each subinterval and computing the Riemann sum.   **pt** may be
chosen as **left**, **right**, **random**, or **any number between 0 and 1**. (If **left(right)** is chosen, the
point taken in each subinterval is the **left(right)**-hand endpoint of the subinterval. If **random** is
chosen, a point in each subinterval is chosen randomly. If **pt** is chosen as a number between 0 and 1,
then the point chosen in each subinterval is the point that "fraction" of the distance from the left-hand
endpoint to the right-hand endpoint.) If **graph** is not included, no graph is drawn. This is not a
standard *Mathematica* command. We have created this command to use in this lab. Your instructor
will make it available to you.

■ COMPUTING RIEMANN SUMS

As just indicated, *Mathematica* can be used to calculate Riemann sums with the command
riemannsum and used to investigate properties of limits of Riemann sums.

EXAMPLE 1. Let g(x)=-x (x-1).

❏ **1.1.** Carefully define g and then graph g on the interval [0,1].

❏ **1.2.** To use **riemannsum** to compute a Riemann sum for g on the interval [0,1] by dividing
[0,1] into five subintervals, **Enter** the following command. Indicate and interpret the results.
riemannsum[g[x],{x,0,1},5,random,graph]

❏ **1.3.** Experiment by changing **random** (in the above command) to **left** and then re-executing.
Then replace **random** successively by **right**, **1/3**, and **1/2**, re-executing the command each time.
State the differences noted, both graphically and in terms of the answer. What does each answer
represent?

❏ **1.4. Enter riemannsum[g[x],{x,0,1},5,left]** and interpret the result. (Recall that **graph** is an option: if **graph** is not included, no graph is shown.) Now use **1/2** instead of **left**. Compare results.

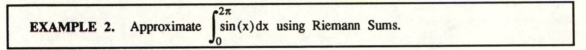

EXAMPLE 2. Approximate $\displaystyle\int_0^{2\pi} \sin(x)\,dx$ using Riemann Sums.

❏ **2.1.** First use *Mathematica* to define f . Then, by **Enter**ing the command **riemannsum[f[x],{x,0,N[2Pi]},25,left,graph]**, we obtain the following result.

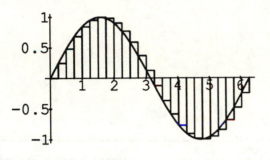

2.45241 10^{-19}

Interpret this result.

❏ **2.2. Experiment** by changing **left** (in the above command) to **right** and then re-executing. Then re-execute the command with **random** substituted for **right**. State the differences noted.

❏ **2.3.** Divide [0,2π] into **50** subintervals and compute a Riemann sum using the **random** option. State the command used and the result. (Do not use the **graph** option.) Repeat the process using **100** subintervals. Compare results.

❏ **2.4.** Deduce the exact value of $\int_0^{2\pi} \sin(x)\,dx$.

▣ PROPERTIES OF INTEGRALS

EXAMPLE 3. Let f(x)=(x-4)(x-6)(x-8) on the interval [0,3].

❏ **3.1.** Explain, algebraically, why f is negative on the interval [0,3].

❏ **3.2.** Compute several Riemann sums on the interval [0,3] using the command **riemannsum**. Use **n**=10, 30, and 50. State the results.

❏ **3.3.** Explain why **riemannsum[f[x],{x,0,3},n,random]** is very close

to $\int_0^3 f(x)\,dx$ for "large" values of n. What can you deduce about $\int_0^3 f(x)\,dx$? If f is a negative continuous function on [a,b], what can you deduce about $\int_a^b f(x)\,dx$?

Area

NAME(S):   INSTRUCTOR:
CLASS: DATE:

▣ INTRODUCTION

The purpose of this lab is to illustrate how *Mathematica* can be used to find the area of a region whose boundaries are determined by the graphs of certain functions. In each case you will be asked to provide two answers to the question,"What is the area of the region bounded by the curves y=f(x) and y=g(x)?", namely, one answer should be given in terms of an integral, or integrals, and the other answer should be a numerical value.

▣ NEW *MATHEMATICA* COMMANDS

There are not really any new *Mathematica* commands in this lab. To refresh your memory, the ones that you might need to use are:

1. **Solve[expression1==expression2,x]**
2. **Solve[{equations},{variables}]**
   For example, **Solve[{2x - 5y == -2,4x + 3y == 5},{x,y}]**
3. **Integrate[f[x],x]** attempts to find an antiderivative of f(x).
4. **Integrate[f[x],{x,a,b}]** attempts to evaluate the definite integral of f with respect to x over the interval [a,b].
5. **NIntegrate[f[x],{x,a,b}]** gives a numerical approximation to the definite integral of f with respect to x over the interval [a,b].
6. **FindRoot[f[x]==g[x],{x,firstguess}]** attempts to solve **f(x)=g(x)** for x starting with the approximation **firstguess**. Remember to try and get a good first guess. Otherwise, **FindRoot** may not be able to find an answer.

▣ FINDING THE AREA OF A REGION

EXAMPLE 1. Consider the region(s) bounded between the curves $f(x) = -x^3 + 3x^2 - 2x$ and $g(x) = -\dfrac{x}{2}$.

First, clear and then carefully define the functions f and g.

❏ **1.1.** Plot the two functions on the interval [-1,3].

❑ LAB # 12 — Areas ❑

The plot of the above two functions shows that there are two regions defined by the intersection of these two curves.

❑ **1.2.** Find the x-values of the intersection points of the curves. State the command(s) used and the x-values obtained.

❑ **1.3.** State the integral that represents the area of the first region and then evaluate it.

HINT. Use Mathematica's notation for representing an integral, namely,

$$\int_a^b f(x)\,dx = \textbf{Integrate[f[x],\{x,a,b\}]}$$

❑ **1.4.** State the integral that represents the area of the second region and then evaluate it.

❑ **1.5.** Find the total area of the regions bounded between the two curves.

EXAMPLE 2. Consider the region bounded between the curves $f(x) = x^4 - 2x^2$ and $g(x) = 2x$.

First, clear and then define the functions f and g.

❑ **2.1.** Plot the functions on the interval [-3,3]. State the command(s) used and sketch the graph.

❑ **2.2.** Find the x-values of the intersection points of the curves. State the command(s) used and the x-values obtained.

❑ **2.3.** What is the integral that represents the area of the region bounded by these two curves?

❑ **2.4.** What is the area of the region bounded by these two curves?

EXAMPLE 3. Consider the region(s) bounded by the two curves x-y+1=0 , 7x - y -17 =0 , and 2x + 2y + 2 = 0.

❑ **3.1.** Plot the three curves to determine the region. State the command(s) used and sketch the graph. HINT. Solve each equation for y to obtain three functions, then plot these functions on the same graph using a different graylevel for each.

❑ **3.2.** Find the x-values of the intersection points of the curves. State the command(s) used and the x-values obtained.

❑ **3.3.** What is the total area of the region bounded by these three curves? Give your answer in both an integral form and a numerical form.

EXAMPLE 4. Consider the region(s) bounded between the two curves $f(x) = \cos(x-2) - \dfrac{1}{2}$

and $g(x) = e^{\left(-(x-2)^2 \cos(\pi(x-2))\right)}$ on $[0,4]$.

❏ **4.1.** By using steps (as in the previous examples), find the total area of the region(s) bounded between these curves. Indicate each step you take to arrive at the final answer. For each step, state the command(s) used and the result(s) obtained. (NOTE. You may need to use NIntegrate to obtain the solution.

EXAMPLE 5. Farmer Brown wants to stake his goat out by the garage so that it will clear some of the grass in that area. The stake is located halfway the length of the garage and 3 feet from the garage as shown below.

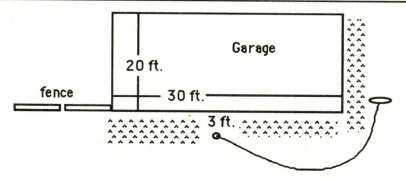

❏ **5.1** How much area can be grazed by the goat if the rope is 20 feet long? Indicate each step you take to arrive at the final answer. For each step, state the command(s) used and the result(s) obtained.

The Fundamental Theorem of Calculus

NAME(S):   INSTRUCTOR:
CLASS: DATE:

▣ INTRODUCTION

THE FUNDAMENTAL THEOREM OF CALCULUS

Let f be a continuous function on an interval $[a, b]$. Define a function G by $G(x) = \int_a^x f(t)\,dt$. Then G is a differentiable function and $G'(x) = f(x)$. If F is any function satisfying $F'(x) = f(x)$, then

$$\int_a^b f(x)\,dx = F(b) - F(a).$$

The purpose of this lab is to illustrate the above properties.

Remark: Depending upon the computer system and version of *Mathematica* you are using, in several cases, the package **IntegralTables.m** must be loaded prior to computing anti-derivatives and definite integrals. If you need to load this package, **Enter** the command **<<IntegralTables.m**

▣ NEW *MATHEMATICA* COMMANDS

As you recall, if $f(x)$ is in a certain class of "simple" functions, then

Integrate [f[x], x] gives the indefinite integral (or antiderivative) $\int f(x)dx$ and

Integrate [f[x], {x, a, b}] gives the definite integral $\int_a^b f(x)dx$.

NIntegrate [f[x], {x, a, b}] numerically approximates the definite integral $\int_a^b f(x)dx$.

NIntegrate[f[x],{x,a,b}] will have to be used instead of **Integrate[f[x],{x,a,b}]** if $f(x)$ is not sufficiently simple. (What do you think happens when one executes this command when that definite integral doesn't exist?)

▪ CAUTION

If F is an antiderivative of f, then *Mathematica* will apply the Fundamental Theorem of Calculus when using the command **Integrate** to evaluate $\int_a^b f(x)\,dx$.

Be careful when you use the Integrate command!! Mathematica does NOT verify that the Fundamental Theorem of Calculus applies. Hence, it is possible to get erroneous results.

▪ EXAMPLES USING THE FUNDAMENTAL THEOREM

EXAMPLE 1. Let $f(x) = \dfrac{1}{x^2}$ on the interval $[-1,1]$.

❑ **1.1.** State the result of **Enter**ing **Integrate[1/x^2,{x,-1,1}]**

❑ **1.2.** By graphing $\dfrac{1}{x^2}$, verify that it is not possible that $\int_{-1}^{1} \dfrac{1}{x^2}\,dx = -2$.

❑ **1.3.** Explain why the Fundamental Theorem of Calculus does not apply.

EXAMPLE 2.   Let $f(x) = \dfrac{x}{2} + \sin(x)$.

❑ **2.1.** Define f and then graph f(x) for $0 \le x \le 2\pi$ to see that f is non-negative on the interval $[0,2\pi]$.

Let z be any number such that $0 \le z \le 2p$ and define $capf(z) = \int_0^z f(x)\,dx$ by **Enter**ing:

Clear[capf]; capf[z_]:= NIntegrate[f[x],{x,0,z}]

NOTE. Since f is non-negative on [0,2π], it follows that capf(z) corresponds to the area between the graph of f, the x-axis, the vertical line x=0, and the vertical line x=z.

❑ **2.2.** Compute the derivative of capf(z). (State the command(s) used and the result.)

❑ **2.3.** Compute and interpret capf(π). (State the command(s) used and the result.)

❑ **2.4.** Compute and interpret capf(2π). (State the command(s) used and the result.)

❑ **2.5.** Graph capf(z) for $0 \leq z \leq 2\pi$. Display the graphs of capf and f simultaneously.

EXAMPLE 3. Let $g(x) = \sqrt{x+3}\ \cos(\pi x)$ on [1,4] and let $capg(t) = \int_1^t g(x)\,dx = \int_1^t \sqrt{x+3}\ \cos(\pi x)\,dx$.

❑ **3.1.** Define g and then graph g on the interval [1,4].

Now, define capg(t). (HINT. Use **NIntegrate**, not **Integrate**. When plotting capg(t) below, the time difference is astounding!)

❑ **3.2.** Compute capg(3/2) and interpret the result.

❑ **3.3.** Compute capg(5/2) and interpret the result.

❑ **3.4.** Save the plot of g(x) on [1,4] as **graph1** using the option **PlotStyle->GrayLevel[0]**

❑ **3.5.** Save the plot of capg(t) on [1,4] as **graph2** using the option **PlotStyle->GrayLevel[.4]** (CAUTION. Make sure you used **NIntegrate** and not **Integrate** in the definition of capg(t)!)

Entering the command **Show[graph1,graph2]** gives the following result. Does it?

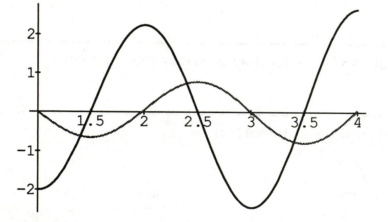

❑ **3.7.** Label the curves **g(t)** and **capg(t)** as appropriate. Then choose (and label) t1,t2,and t3 where capg(t1)<0, capg(t2)=0 and capg(t3)>0. Explain these (inequalities, equality) in terms of areas.

❑ **3.8.** Using the graph as a guideline, it appears that capg is a differentiable function. Find its derivative.

The Mean Value Theorem For Integrals

NAME(S):   INSTRUCTOR:
CLASS: DATE:

▪ INTRODUCTION

The purpose of this lab is to examine the **Mean-Value Theorem for Integrals** and to calculate the **average** (or **mean**) **value** of an integrable function on the interval [a,b].

The **Mean - Value Theorem for Integrals** states: If f is continuous on [a, b], then there is at least one number c in [a, b] such that $\int_a^b f(x)dx = f(c)(b - a)$. Dividing both sides by b - a, it follows that there is at least one number c in [a, b] such that $f(c) = \frac{1}{b-a}\int_a^b f(x)dx$.

If f is integrable on [a, b], then the **average value** (or **mean - value**) of f over [a, b] is defined to be $f(x)_{ave} = \frac{1}{b-a}\int_a^b f(x)$.

It now follows that if f is continuous on [a,b], then there is at least one number c in [a,b] such that f(c) is the average value of f over [a,b].

▪ NEW *MATHEMATICA* COMMANDS AND PROCEDURES

FindRoot[exp1==exp2,{variable,first guess}] attempts to numerically approximate solution(s) to the equation **exp1=exp2** by finding roots via Newton's Method. Hence, your first guess needs to be a point where the derivative of both sides of the equation is defined. You should also be aware that Newton's Method doesn't always work and, even when it does, your first guess needs to be "close to" the actual root. A plot of **exp1-exp2** will normally enable one to obtain a good first guess since the x-coordinates of the points where the graph of **exp1-exp2** intersects the x-axis correspond to the solutions of the equation **exp1=exp2**. **FindRoot** is typically used when the **Solve** command isn't able to find the desired solution(s).

randomtriangle[length1, length2] shows a triangle with two sides of lengths **length1** and **length2**, and third side of random length between **length1+length2** and **|length1-length2|**, and then uses Heron's Formula to compute its area. This is not a standard *Mathematica* command but was created by us for use in this lab. It is available from your instructor.

■ USING FINDROOT TO SOLVE EQUATIONS

EXAMPLE 1. Solve the equation $\sin\sqrt{x} = \sqrt{\cos(x)}$ on the interval $[0,\frac{\pi}{2}]$.

First, put the equation in the form f(x)=0 and define the function f(x) by **Enter**ing the following commands.
Clear[f]
f[x_]=Sin[Sqrt[x]]-Sqrt[Cos[x]]

❏ **1.1.** Try to find a solution of f(x)=0 by using the **Solve** command. What is your result?

❏ **1.2.** Now let's find a solution by using the **FindRoot** command. Recall that the first approximation needs to be a good one, so let's first plot f(x) on the given interval by **Enter**ing the command **Plot[f[x],{x,0,Pi/2}]**

❏ **1.3.** Based on this graph, choose a good first guess. (**first guess=**_____)
Put this value in the following command for **first guess** and then **Enter**
ans=FindRoot[f[x]==0,{x,first guess}]

ans=

❏ **1.4.** Use the *Mathematica* command **N** to obtain the value of your answer to 10 decimal places.

ans=

EXAMPLE 2. Let $h(x) = \sqrt[3]{1 + \sin^3(x^2)}$.

❏ **2.1.** Find all values of x in [0,3] such that h(x)=.6. List the commands you enter and your answer.

■ THE MEAN-VALUE OF A FUNCTION

EXAMPLE 3. Find the average value of $f(x) = (2x+1)^2$ on the interval $[-1,1]$.

The average value of f(x) on the interval $[-1,1]$ is given by:

$$f(x)_{ave} = \frac{\int_{-1}^{1} (2x+1)^2 \, dx}{1-(-1)} = \frac{\int_{-1}^{1} (2x+1)^2 \, dx}{2}.$$

First, define the function f.

❏ **3.1.** Find the average value by **Entering avevalue=Integrate[f[x],{x,-1,1}]/2**

avevalue=

❏ **3.2.** Find all numbers c that satisfy the conclusion of the Mean Value Theorem for Integrals for f(x) on [-1,1]. In order to answer this, we need to solve the equation **f(c)=avevalue** for all c in [-1,1]. First try using the **Solve** command. If that is not successful, then use the **FindRoot** command. Make sure your answer is correct to 10 decimal places. **Enter**

roots=Solve[f[c]==avevalue,c]

roots=

❏ **3.3.** Which of the above solutions is a valid solution to the problem? Why?

EXAMPLE 4. Let $g(x) = \cos\sqrt{x} - \sqrt{\sin(x)}$.

❏ **4.1.** Does the Mean Value Theorem for Integrals apply to g(x) on $[0,\pi]$? Why or why not? on $[0,2\pi]$? Why or why not?

❏ **4.2.** If the theorem applies to g(x) on $[0,\pi]$, find the average value of g(x) over $[0,\pi]$, and all values of c that satisfy the theorem. List the commands you enter and your answers.

EXAMPLE 5. Find a number w such that $\left| \int_{1}^{w} \frac{1}{x} \, dx - 1 \right| \le .005$.

First, let's define $f(w) = \int_{1}^{w} \frac{1}{x} dx$. To do this, **Enter** the following.

71

Clear[f]; f[w_]:=NIntegrate[1/x,{x,1,w}]

❏ **5.1.** Now, plot f(w)-1 over [1,3] by **Entering Plot[f[w]-1,{w,1,3}]** to get a "feel" for the function.

❏ **5.2.** By "squeezing down" about a zero of f(w)-1, find w as specified above.

    w=

EXAMPLE 6. Let T denote triangle ABC with AB=8 and BC=11. What must the length of CA be in order to yield the average value of the area of T?

Recall that, in general, the area of triangle T with sides of length a, b, and c is given by **Heron's Formula:**

Area of triangle $T = \sqrt{s(s-a)(s-b)(s-c)}$; where $s = \dfrac{1}{2}(a+b+c)$.

Let x denote the length of the third side. Then x must be bigger than 3 and less than 19. By Heron's Formula, the area of T as a function of x is given by:

$$\text{area}(x) = \sqrt{\left(\frac{x+8+11}{2}\right)\left(\frac{x+8+11}{2}-x\right)\left(\frac{x+8+11}{2}-8\right)\left(\frac{x+8+11}{2}-11\right)} \ .$$

Let's use *Mathematica* to define area(x) on [3,19]:
Clear[area]
area[x_]:=Sqrt[(x+8+11)/2   ((x+8+11)/2-x)   ((x+8+11)/2-8)   ((x+8+11)/2-11)]

❏ **6.1.** Let's examine the area of randomly generated triangles with one side of length 8 and one side of length 11. State the result of **Entering randomtriangle[8,11]**

❏ **6.2.** By repeating the above experiment five times and averaging your results, approximate the average area of triangles that have one side of length 8 and another of length 11. (Indicate the five individual areas and the average.)

❏ **6.3.** Compute the average value of area(x) on the interval [3,19].

    avevalue=

❏ **6.4.** Finally, find the length of the third side that gives this average area. State the commands you enter and the answer.

Volumes of Solids of Revolution

NAME(S):   **INSTRUCTOR:**
CLASS: **DATE:**

▣ INTRODUCTION

The purpose of this lab is to calculate the volumes of solids of revolution. In order to help you understand, we will do some three-dimensional graphics.

Let f(x) be a non-negative continuous function on an interval [a,b], with a >0. Let R denote the region bounded by the x-axis, and the graphs of y=f(x), x=a, and x=b. The **volume of the solid generated by revolving R around the x-axis** is given by:

$$V_{xaxis} = \int_a^b \pi (f(x))^2 \, dx \; .$$

The **volume of the solid generated by revolving R around the y-axis** is given by:

$$V_{yaxis} = \int_a^b 2\pi x \, f(x) \, dx \; .$$

▣ IMPORTANT NOTE ABOUT MEMORY *(**READ THIS**)*

The purpose of this note is to make you aware of some of the limitations of *Mathematica* and attempt to save you from "crashes" and heartaches later.

Three-dimensional graphics use a **very large** amount of the computer's memory. On a Macintosh, the amount of memory you have used during a *Mathematica* session is displayed at the bottom of the screen (the thermometer). The white area represents the amount of memory you have available. Beware: when the thermometer "gets full", save your file, then quit your current *Mathematica* session and restart. (If your computer does not have a thermometer, be very careful not to try to do too many graphs before saving your file, quitting your session, and then restarting.) *After you restart, be sure to re-execute the definitions of the functions you need.*

▣ NEW *MATHEMATICA* COMMANDS

1. **solidrev[f[x],{x,a,b},axis]** yields a three-dimensional meshed image of the function f(x) defined on the domain [a,b] revolved about the x-axis or y-axis.

2. **solidrev[f[x],{x,a,b},axis,solid]** yields a solid surface. The interval [a,b] is automatically

divided into 10 subintervals. This may be changed by substituting {x,a,b},n for {x,a,b} where n is the desired number of subintervals.

solidrev is not a standard Mathematica function. It has been written for use in this lab and is available from your instructor.

The following are not new but will be listed for your convenience:

Together[expression]
FindRoot[f[x]==g[x],{x,firstguess}]
ExpandDenominator[expression]
NIntegrate[f[x],{x,a,b}]
Integrate[f[x],x]
Integrate[f[x],{x,a,b}]

■ VOLUMES OF SOLIDS OF REVOLUTION

EXAMPLE 1. Let $f(x) = 2x^2 - 2x^3$ and let R denote the region bounded by the x – axis, $y = f(x)$, $x = 0$, and $x = 1$.

First , define f.

❏ **1.1.** Plot f(x) on [0,1].

❏ **1.2.** Calculate the volume of the solid obtained by revolving R about the x-axis. State the command(s) used and the result(s).

❏ **1.3.** Visualize the solid by **Entering solidrev[f[x],{x,0,1},xaxis,solid]**

❏ **1.4.** Calculate the volume of the solid obtained by revolving R about the y-axis. State the command(s) used and the result(s).

❑ **1.5.** Visualize the solid by **Entering solidrev[f[x],{x,0,1},yaxis,solid]**

Notice that the volumes in 1.2 and 1.4 are different!

EXAMPLE 2. Let $0 < k < 1$ and let $h_k(x) = (1-k)\cos\left(\dfrac{x}{k}\right)$.

❑ **2.1.** Verify, algebraically and geometrically, that $h_k(x)$ is non – negative and continuous

on $\left[0, \dfrac{k\pi}{2}\right]$.

Let V_k denote the volume of the solid obtained by revolving the region bounded by the graphs

of $h_k(x)$, $x = 0$, $x = \dfrac{k\pi}{2}$, and the x - axis around the y - axis. Define $h_k(x)$ by **Entering** the

following commands.

Clear[h]
h[x_,k_]=(1-k) Cos[x/k]

❑ **2.2.** Now, just out of curiosity, let's plot the function defined by k=.5. This is accomplished by
Entering Plot[h[x,.5],{x,0,Pi/4}]

Then the volume, V_k, of the solid obtained by revolving the region bounded by the graphs of $h_k(x)$, $x = 0$, $x = \dfrac{k\pi}{2}$, and the x‑axis around the y‑axis is given by

$$V_k = \int_0^{\frac{k\pi}{2}} 2\pi x\, h_k(x)\, dx = \int_0^{\frac{k\pi}{2}} 2\pi x\, (1-k)\cos\left(\frac{x}{k}\right) dx.$$

Now define V(k) by **Entering** the following commands.
Clear[V,k]
V[k_]:= Integrate[2 Pi x (1-k) Cos[x/k],{x,0,(k Pi/2)}]

❏ **2.3.** To see that V(k) attains a maximum, graph V(k) on [0,1].

❏ **2.4.** Now find the maximum value by locating where the derivative is zero and evaluating **V** there.

❏ **2.5.** Now, visualize the solid by entering the following commands.
Clear[f]
f[x_]:=h[x,2/3]
solidrev[f[x],{x,0,N[Pi/3]},yaxis,solid]

Work = Force through Distance

NAME(S):   **INSTRUCTOR:**
CLASS: **DATE:**

▪ INTRODUCTION

Work is done on an object when a force is used to move the object from point A to point B. If the force is constant and the object moves along a line in the same direction as the applied force, then the work done in moving the object from point A to point B is just the product **force** times **distance from A to B**. If the force is not constant along the path AB or if the distance of the applied force is varying, then integral calculus must be used to obtain the amount of work done in moving an object from A to B.

There are two fundamental prototypes involved in the calculation of work. They are:

1. A type that arises from an **extrinsic force** applied to the **same object** as it moves from point A to point B.

   This is the type of situation that arises when a rocket is being lifted from the earth's surface to a certain height. A force is being applied to the rocket externally, but the force varies at each point X between point A, the earth's surface, and point B. We can make use of the formula **Work = (Force) * (Distance)** by partitioning the interval $[A,B]$ into finitely many small intervals $[x_i, x_{i+1}]$. Assuming the force function is continuous, we can assume the force function is constant over each interval $[x_i, x_{i+1}]$. Using this assumption, we get the following approximation:

$$\text{Work} \overset{\text{approx.}}{\approx} \sum_i (\text{Work moving the object from point } x_i \text{ to point } x_{i+1})$$

$$\overset{\text{approx.}}{\approx} \sum_i (\text{Force at } x_i) * \Delta x_i .$$

   Upon taking the limit as Δx_i goes to 0, we get $\text{Work} = \lim_{\Delta x_i \to 0} \sum_i (\text{Force at } x_i) * \Delta x_i$.

   The above sum is a Riemann sum, and thus the work is represented as an integral.

2. A type that arises from an **intrinsic force** that is being applied to **different slices** of an object. This situation arises when trying to calculate the work done in emptying a tank full of water. In this case, one wants to calculate the work done in lifting each slice of water to the top of the tank. Therefore, we can view the total work as:

Work $\overset{approx.}{\approx} \sum_i$ (Work to lift the ith slice from position x$_i$ to point B)

$$= \sum_i (\text{Mass of i}^{th}\text{slice}) * (\text{distance the i}^{th}\text{slice moves})$$

Again, taking limits as the width of the ith slice goes to 0, and using the fact that mass = density * volume, we get Work =

$$\lim_{\substack{\text{width i}^{th} \\ \text{slice} \to 0}} \sum_i (\text{Density of the i}^{th}\text{slice})*(\text{Volume of the i}^{th} \text{ slice})* (\text{distance the i}^{th} \text{ slice moves}).$$

Since the above sum will be a Riemann sum, our answer will be expressed in terms of an integral.

▣ NEW *MATHEMATICA* COMMANDS

There are no new *Mathematica* commands used in this lab.

▣ WORK DUE TO EXTRINSIC FORCES

> **EXAMPLE 1. Spring problem.** According to Hooke's law, the force required to stretch a spring x units beyond its natural length is given by force(x) = kx. Suppose a force of 9 pounds is required to stretch a spring from its natural length of 6 inches to a length of 8 inches.

❏ **1.1.** Find the value of k, the spring constant.

❏ **1.2.** Find the work done in stretching a spring from its natural length, 6 inches, to 10 inches.

Partition the interval [6,10] into a finite number of points x$_i$. Then ask the questions:

1) What is the force at x$_i$?

2) What is the integral that represents the work done in stretching the spring from 6 inches to 10 inches?.

See the Riemann sum suggested in the introduction.

EXAMPLE 2. Leaking bucket. A bucket containing water is lifted vertically at a constant rate of 1.5 ft/sec by means of a rope of negligible weight. As the bucket rises, water leaks out at the rate of 0.25 lb/sec. If the bucket weighs 4 pounds when empty and if the bucket originally contained 20 pounds of water, determine the work done in raising the bucket h feet.

Partition the interval $[0,8]$, that represents the height, into a finite number of intervals x_i.

❏ **2.1.** How much water has leaked out of the bucket when the bucket reaches the point x_i?

❏ **2.2.** How much work is done in lifting the bucket 8 feet?

❏ **2.3.** How much work is done in lifting the bucket 12 feet? Be careful, this is tricky.

HINT. Refer back to the Riemann sum suggested in the introduction.

EXAMPLE 3. Rocket.

Newton's law of gravitation states that the force F of attraction between two particles m_1 and m_2 is given by $F = \dfrac{G m_1 m_2}{d^2}$, where G is a gravitational constant and d is the distance between the particles. Suppose the mass m_1 of the earth is regarded as concentrated at the center of the earth and a rocket of mass m_2 is on the surface (a distance of 4000 miles) and is fired vertically upward to a height of H miles.

❏ **3.1.** Partition the height $[0,H]$ into a finite number of points h_i. What is the force on the rocket when it reaches the point h_i?

79

❑ **3.2.** Find the work done in firing the rocket vertically upward to an altitude of H miles.

▪ WORK DUE TO AN INTRINSIC FORCE

> **EXAMPLE 4. Emptying a container.** A hemispherical water tank has radius 5 feet. The tank is mounted with its circular base on top, lying horizontally. The tank is emptied by pumping the water out of a valve located on the top edge of the tank.

Let the variable h denote the height. Partition the interval [0,5] and imagine slicing the water volume by horizontal plates into slabs of thickness Δh_i. Recall that the density of water is

62.4 lb/ft^3.

❑ **4.1.** What is the volume of the i-th slab?

❑ **4.2.** How much work is required to lift the i-th slab out of the water?

❑ **4.3.** How much work is required to pump all the water out of the tank if the tank is full?

EXAMPLE 5. Lifting a chain. Suppose that a ship's anchor weighs 2 tons (4000 pounds) in water and that the anchor is hanging taut from 100 feet of cable. Assume the cable weighs 20 pounds per foot in water.

Partition the interval [0,100] into a finite number of points h_i.

❏ **5.1.** Approximately how much work is required to lift the i-th section of cable to the surface? HINT. Refer back to the Riemann sum suggested in the introduction for **type 2** problems.

❏ **5.2.** How much work is required to raise the cable?

❏ **5.3.** How much work is required to raise the anchor to the surface of the water?

EXAMPLE 6. Work required in emptying a coke can.

❑ **6.1.** If the can is vertical, find the amount of work required to remove all the liquid from a can of coke (from the top). Assume the can has dimensions as shown in Figure A below.

Figure A

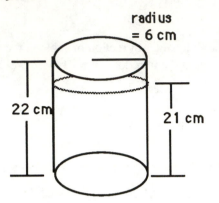

radius
= 6 cm

22 cm

21 cm

height of can = 22 cm

height of liquid in the can
= 21 cm

❑ **6.2.** If the can is horizontal (see Fig. B), find the amount of work required to empty the can (through the hole indicated on the "top"). Would you expect it to be the same? Why or why not? Is it the same? Explain.

Figure B

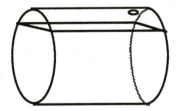

Inverse Functions

NAME(S):   **INSTRUCTOR:**
CLASS: **DATE:**

■ INTRODUCTION

The purpose of this lab is to study the very important topic of inverse functions. As you recall, a function has an inverse if and only if it is a one - to - one function. If a function f is one - to - one, then its inverse is denoted by f^{-1}. f and f^{-1} satisfy the property that $y = f(x)$ if and only if $x = f^{-1}(y)$. It follows that $f(f^{-1}(x)) = x$ for every x in the domain of f^{-1} (which is the range of f) and $f^{-1}(f(x)) = x$ for every x in the domain of f (which is also the range of f^{-1}).

RECALL. A function f(x) is **one-to-one** (also denoted **1-1**) means that if x is not equal to y, then f(x) is not equal to f(y). In terms of the graph of f, this means that every horizontal line passes through the graph of f at most one time. Consequently, every strictly increasing function is one-to-one and every strictly decreasing function is one-to-one.

Thus, there are two tests that can be used to determine whether or not f(x) has an inverse (or, **f(x) is invertible**) on [a,b].

Test 1. Geometric Test
        If no horizontal line cuts the graph of f in more than one point
        then f is one-to-one on [a,b] and thus f has an inverse on [a,b].

Test 2. Derivative Test
        If f is differentiable on [a,b] and f' has the same sign throughout (a,b)
        then f is one-to-one on [a,b] and hence f is invertible on [a,b].

■ NEW *MATHEMATICA* COMMANDS AND PROCEDURES

The new *Mathematica* command used in this lab is **inversegraph** .

Let f(x) be a function on [a,b]. Observe that the graph of f(x) on [a,b] is the set of points F={(x,f(x)): x in [a,b]}. Let G={(f(x),x): x in [a,b]}. We will call G the **inverse graph of f(x)**. Note that G does not necessarily represent a function g. (Or, using the ordered pairs definition, G is not necessarily a function.) G will represent a function g if-and-only-if f is 1-1.

inversegraph[f[x],{x,a,b}] plots **f(x)**, the **inverse graph of f(x)** (in gray), and the line **y=x** (dashed). This is not a standard *Mathematica* command. It has been created for use in this lab and is available from your instructor.

There are no other new *Mathematica* commands in this lab but we will mention several in discussing some capabilities that will be especially useful in this lab.

Solve for x in terms of y
Suppose we have a function f(x) and we want to solve y=f(x) for x in terms of y. Then, after defining the function f, **Enter** the following.
Solve[f[x]==y,x]

Simplifying expressions
Suppose we want **to rewrite a sum of rational expressions as a single rational expression.** Then use the **Together** command as shown in the following.

exp1=1/x+2x/(x-3)
Together[exp1]
exp2=(Exp[x] - Exp[-x]) / (Exp[x]+Exp[-x])
Together[exp2]

To simplify an algebraic expression, use the **Simplify** command as follows.

exp3=(x^3-5x^2+6x)/(2x^3-5x^2+2x)
Simplify[exp3]

▪ INTRODUCTION TO INVERSE FUNCTIONS

> **EXAMPLE 1.** Let $f(x) = (x-2)^2 - 1$ on [0,4].

Using **inversegraph**, let's graph both f(x) and its inverse graph. First define f by **Enter**ing the commands **Clear[f]; f[x_]:=(x-2)^2 - 1**
Then, **Enter**ing the command **inversegraph[f[x],{x,0,4}]** yields the following graph.

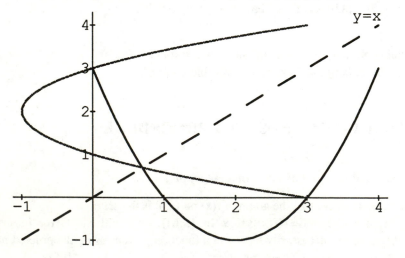

❏ **1.1.** Using **Test 1**, does f have an inverse on [0,4]? Why or why not?

❑ **1.2.** If f is not 1-1 on [0,4], is f 1-1 on some subinterval of [0,4]? If so, find such a subinterval of maximum width and plot both f and its inverse on that subinterval.

EXAMPLE 2. Restricting the domain of f(x) so that f will be invertible over the restricted domain.

For each of the following functions find the largest interval, containing the point x_0, on which f is invertible. (HINT. Find f'(x) and find where it is 0, +, -. Then use Test 2.) In each case, specify the domain of f^{-1}.

❑ **2.1.** $f(x) = x^2$, $x_0 = -3$

❑ **2.2.** $f(x) = x^3 - 3x^2 + x$, $x_0 = -1$

EXAMPLE 3. Let $g(x) = \sin(x)$ on $[-6\pi, 6\pi]$.

❑ **3.1.** Using Test 1 or Test 2, does g have an inverse on $[-6\pi, 6\pi]$? Why or why not?

❑ **3.2.** If g does not have an inverse on $[-6\pi, 6\pi]$, then find an interval containing 0 having largest width and such that g does have an inverse on this interval. State the interval and then use inversegraph to display the graphs of g and g^{-1}. (NOTE. $g(x) = \sin(x)$ is invertible on this interval and its inverse is $g^{-1}(x) = \sin^{-1}(x) = \arcsin(x)$.)

❑ **3.3.** State the domain and range of arcsin(x).

❑ **3.4.** Define h(x) by the commands **Clear[h]; h[x_]:=ArcSin[x]** and then plot h(x) on its domain. How does this compare to the graph of the inverse of sin(x) above?

EXAMPLE 4. Let g(x)=tan(x).

❑ **4.1.** Plot g(x) on a large interval and determine whether on not g is invertible.

❑ **4.2.** Find the largest interval containing 0 on which g is invertible. Then plot g(x) and its inverse.

❏ **4.3.** What is the name of the inverse function? State its domain and range. What command does *Mathematica* use to define it?

EXAMPLE 5. Finding the inverse and verifying this result.

One ability that is important is the ability to solve the equation $y = f(x)$ for x in terms of y (i.e. to express x as a function of y ($x = g(y)$)). This can be done if and only if $f(x)$ is one-to-one (for each x there is one and only one y such that $y = f(x)$). If this is the case then we know that $g(f(x))=x$ for all x in the domain of f and $f(g(y))=y$ for all y in the range of f.

In the following, let $f(x) = \dfrac{2x - 1}{3x + 1}$. First, carefully define the function f and verify that it is 1 - 1.

❏ **5.1.** Using the **Solve** command, solve the equation $y=f(x)$ for x. (Note from above that $x=g(y)$, where g is the inverse of f.)

$g(y) =$

❏ **5.2.** Verify that $g(f(x))=x$ for all x in the domain of f. (State the domain.)

❏ **5.3.** Verify that $f(g(y))=y$ for all y in the range of f. (State the range.)

❏ **5.4.** State the domain of g and the range of g.

EXAMPLE 6. Let $f(x) = \displaystyle\int_0^x \sqrt{1 + t^4}\ dt$ for all x.

First, carefully define f using the **NIntegrate** command.

❏ **6.1.** Using **inversegraph**, plot f and its inverse graph. Does f appear to have an inverse?

❏ **6.2.** Verify that f has an inverse by Test 2.

EXAMPLE 7. A photographer is taking a picture of a five-foot painting hung in an art gallery. The camera lens is one foot below the lower edge of the painting.

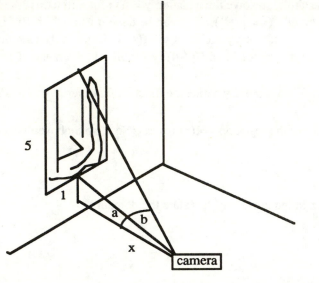

❏ **7.1.** How far should the camera be from the painting to maximize the angle subtended by the lense? (Notice that b is the angle we wish to maximize; x is the distance from the wall to the camera.)

Logarithmic / Exponential Functions & Rates of Growth

NAME(S):   INSTRUCTOR:
CLASS: DATE:

▨ INTRODUCTION

The purpose of this lab is to examine exponential functions, logarithmic functions, and their rates of growth compared to polynomials. In addition to examining graphs of these functions, we will examine the graphs of $\dfrac{\ln(x)}{q(x)}$, $\dfrac{q(x)}{\ln(x)}$, $\dfrac{e^x}{q(x)}$, and $\dfrac{q(x)}{e^x}$, where q(x) is any polynomial with degree greater than or equal to 1.

▨ NEW *MATHEMATICA* COMMANDS AND PROCEDURES

1. **randompoly[deg]** generates a polynomial p(x) with degree **deg** and random integer coefficients between -100 and 100. **randompoly[deg,n]** generates a polynomial p(x) with degree **deg** and random integer coefficients between **-n** and **n**. This is not a standard *Mathematica* command. It was created for use in this lab and is available from your instructor.

2. **randompolynomial** generates a random polynomial p(x) with degree less than 50 and integer coefficients between -100 and 100. The definition for **randompolynomial** is in the initialization cell. This is not a standard *Mathematica* command. It was created for use in this lab and is available from your instructor.

3. **Limit[f[x],x−>a]** attempts to compute $\underset{x\to a}{\text{Lim}}\, f(x)$.

4. **Log[x]** gives the **natural logarithm** of x. **Log[b, z]** gives the logarithm to base b.
   Remember: In the literature, ln (x) denotes the natural logarithm of x. But, for our purposes, whenever we refer to **Log[x]**, this means "the natural logarithm of x."

5. **Exp[x]** computes e^x, where e is the UNIQUE number satisfying $\displaystyle\int_1^e \frac{1}{t}\,dt = 1$.

89

6. The command **inversegraph** was used in the previous lab but will also be used in this lab.

Let f(x) be a function on [a,b]. Observe that the graph of f(x) on [a,b] is the set of points F={(x,f(x)): x in [a,b]}. Let G={(f(x),x): x in [a,b]}. We will call G the **inverse graph of f(x)**. Note that G does not necessarily represent a function g. (Or, using the ordered pairs definition, G is not necessarily a function.) G will represent a function g if-and-only-if f is 1-1.

inversegraph[f[x],{x,a,b}] plots f(x), the **inverse graph of f(x)** (in gray), and the line y=x (dashed). This is not a standard *Mathematica* command. It has been created for use in this lab and is available from your instructor.

▪ LOGARITHMIC FUNCTIONS

> **EXAMPLE 1.** Consider ln(x), ln(x+k), and ln(x)+k.

Let **graph1** be the plot of ln(x) on [.01,5], **graph2** be the plot of ln(x+2) on [-1.99,5], and **graph3** be the plot of ln(x)+2 on [.01,5].

❏ **1.1. Show** all three graphs on one set of axes using different **GrayLevels** so one can "pick out" the different functions. Label the graphs and indicate how the graphs compare to each other.

❏ **1.2.** Display the graphs of ln(x), ln(x-5), and ln(x)-5 on one set of axes using different **GrayLevels** so one can "pick out" the different functions. Label the graphs and indicate how the graphs compare to each other.

❏ **1.3.** Without graphing, compare the graphs of ln(x), ln(x+k), and ln(x)+k.

> **EXAMPLE 2.** Consider the graphs of $\ln(x)$, $\ln(kx)$, and $\ln\left(\dfrac{x}{k}\right)$ where $k > 0$.

❏ **2.1.** Without graphing, compare the graphs of these functions. HINT. Use properties of logarithms and then refer to EXAMPLE 1 above.

■ EXPONENTIAL FUNCTIONS

> **EXAMPLE 3.** Consider the graphs of e^x, e^{x+k}, and $e^x + k$.

❏ **3.1.** For k=-1, show the graphs of all three functions on [-5,5]. (Use different GrayLevels.) How do they compare?

❏ **3.2.** For arbitrary k, how do the three functions compare?

> **EXAMPLE 4.** Consider e^x, e^{kx}, and $e^{\frac{x}{k}}$ for arbitrary k.

❏ **4.1.** For k=2, show the graphs of all three functions on [-5,5]. (Use different GrayLevels.) How do they compare?

❑ **4.2.** For arbitrary k, how do the three functions compare?

◼ RELATIONSHIP BETWEEN LOG AND EXP FUNCTIONS

EXAMPLE 5. Relationship between the graphs of the Log and Exp functions

❑ **5.1.** Use inversegraph to graph Exp[x] and its inverse graph for x in [-4,3].

❑ **5.2.** Use inversegraph to graph Log[x] and its inverse graph for x in [.01,20].

❑ **5.3.** Compare the two results above and state your observations.

As you recall, a function has an inverse if and only if it is a one - to - one function. If a function f is one - to - one, then its inverse is denoted by f^{-1}. f and f^{-1} satisfy the property that y = f(x) if and only if x = f^{-1}(y). It follows that f(f^{-1}(x)) = x for every x in the domain of f^{-1} (which is the range of f) and f^{-1}(f(x)) = x for every x in the domain of f (which is also the range of f^{-1}).

❏ **5.4.** Verify that Exp[x] and Log[x] are inverses of each other. State the domain and range of each.

▨ LOGARITHMIC FUNCTIONS DIVIDED BY POLYNOMIALS

> **EXAMPLE 6.** Let q(x) be any polynomial. Compute $\lim\limits_{x\to\infty} \dfrac{\ln(x)}{q(x)}$.

Experimental Approach

Generate a random polynomial p(x) by **Entering** the command **randompolynomial**

❏ **6.1.** Enter the command **?p** and explain the result. Compare your result with your classmates.

❏ **6.2.** Plot the function $\dfrac{\ln(x)}{p(x)}$ on [ε,N] where ε is a SMALL positive number, like .001, (since ln(0) is undefined) and N is a LARGE positive number, like 1000, by **Entering Plot[Log[x] / p[x],{x,.001,1000}]**

❏ **6.3.** Modify the above command by replacing 1000 by numbers MUCH bigger than 1000 and predict $\lim\limits_{x\to\infty} \dfrac{\ln(x)}{p(x)}$. State the numbers you chose and why you reached the conclusion you did.

❑ **6.4.** Compare your answer with the result of **Enter**ing the command
   Limit[Log[x]/p[x], x->Infinity]

❑ **6.5.** Try the same procedure again by **Enter**ing the following two commands.
   randompolynomial
   Plot[Log[x]/p[x],{x,10000,100000}]

❑ **6.6.** In this case, predict $\displaystyle\lim_{x\to\infty}\frac{\ln(x)}{p(x)}$.

❑ **6.7.** If q(x) is ANY polynomial with degree greater than or equal to one, calculate $\displaystyle\lim_{x\to\infty}\frac{\ln(x)}{q(x)}$.

EXAMPLE 7. Sketch the graph of each of the following functions on the indicated interval. Find and classify all relative extrema and points of inflection.

❑ **7.1.** $f(x) = 1 - e^{-x^2}$, Interval $(-\infty.\infty)$

❑ **7.2.** $g(x) = e^x \cos(x)$, Interval $[-2\pi, 4\pi]$

94

Improper Integrals

NAME(S):   INSTRUCTOR:
CLASS: DATE:

▣ INTRODUCTION

Many integrals that arise in practice are integrals that require integration over an interval of infinite

length such as $\int_0^\infty f(x)\,dx$ or integrals on an interval where the function is discontinuous such

as $\int_0^1 \frac{1}{x-1}\,dx$ or $\int_{-2}^2 \frac{1}{x^2}\,dx$. Integrals of this type are called **improper integrals**.

We can define these integrals in terms of limits. For a particular integral, if the limit(s) exists, we assign a value to that integral. The purpose of this lab is to investigate integrals of this type.

▣ NEW *MATHEMATICA* COMMANDS

The main commands used in this lab are **Integrate**, **Limit** and **Plot**, commands which you have already used. In the following examples, you will use *Mathematica*'s ability to evaluate integrals and limits to quickly evaluate improper integrals. Some examples of how we can set up integrals of this type are:

Clear[a]
ans1=Integrate[1/Sqrt[x] , {x,a,1}]   (* What type of improper integral is this?*)
Limit[ans1,a->0]

Clear[b]
ans2=Integrate[1/x^2,{x,1,b}]   (* What type of improper integral is this?*)
Limit[ans2,b->Infinity]

▣ EVALUATION OF IMPROPER INTEGRALS

> **EXAMPLE 1.** Consider $\int_0^\infty t^m\, e^{-t}\, dt$.

❏ **1.1.** Evaluate the above integral for m=1,2,3, and 4.

❏ **1.2.** Plot simultaneously each function $f(t) = t^m e^{-t}$, $m = 1, 2, 3$, and 4, on the interval $[0, 50]$. Choose an appropriate value for the **PlotRange** option in order to plot the entire function.

EXAMPLE 2. Consider $f(x) = \dfrac{1}{(x-1)}$ and $g(x) = \dfrac{1}{(x-1)^2}$ for $x \geq 2$.

❏ **2.1.** Plot simultaneously the functions f(x) and g(x) for $x \geq 2$.

Let **Rf** be the region below y=f(x), above y=0, and to the right of x=2.

❏ **2.2.** Calculate the area **areaf** of **Rf**. State the command(s) used as well as your result.

❏ **2.3.** Calculate the volume **volf** obtained by revolving **Rf** about the x-axis. State the command(s) used as well as your result.

❏ **2.4.** **Intuitively** explain (and reconcile) the results obtained for **areaf** and **volf** above.

Let **Rg** be the region below y=g(x), above y=0, and to the right of x=2.

❑ **2.5.** Calculate the area **areag** of **Rg**. State the command(s) used as well as your result.

❑ **2.6.** Calculate the volume **volg** obtained by revolving **Rg** about the x-axis. State the command(s) used as well as your result.

EXAMPLE 3. The **gamma function** $\Gamma(n) = \int_0^\infty t^{n-1}e^{-t}\,dt$.

Integrals of the form $\int_0^\infty f(x)\,dx$ often arise from attempts to answer questions about the long−term behavior of some process (which is defined by the integral). A typical integral that arises in many cases is the **gamma** function $\Gamma(\mathbf{n})$.

This function is defined in *Mathematica* as **Gamma(t)**.

❑ **3.1.** Plot $\Gamma(t)$ (**Gamma**(t)) on the interval [0,7].

❑ **3.2.** Use integration by parts to show that $\Gamma(t+1) = t\,\Gamma(t)$.

❑ **3.3.** Using the formula established above, find $\Gamma(2), \Gamma(3), \Gamma(4),$ and $\Gamma(10)$.
     HINT: What is $\Gamma(1)$?

❏ LAB # 19 — Improper Integrals ❏

Another type of uncertainty in calculating an integral arises because the function has a discontinuity in the interval of integration. However, by making use of limits, we can frequently assign a value to the integral (provided the limit exists).

EXAMPLE 4. Let $f(x) = \dfrac{1}{\sqrt{4-x}}$ and $g(x) = \dfrac{1}{(4-x)^2}$.

❏ **4.1.** Plot simultaneously f(x) and g(x) over the interval [0,3.9].

❏ **4.2.** Consider the integral $\displaystyle\int_0^4 f(x)\,dx$. Why is it an improper integral? Determine whether or not it converges. If it converges, find its value. State the command(s) you use as well as the result.

❏ **4.3.** Consider the integral $\displaystyle\int_0^4 g(x)\,dx$. Why is it an improper integral? Determine whether or not it converges. If it converges, find its value. State the command(s) you use as well as the result.

❏ **4.4.** Consider the integral $\displaystyle\int_0^8 g(x)\,dx$. Why is it an improper integral? Determine whether or not it converges. If it converges, find its value. State the command(s) you use as well as the result.

Monotone Sequences and Rates of Growth

NAME(S):   INSTRUCTOR:
CLASS: DATE:

▪ INTRODUCTION

The objective of this lab session is to investigate the relative rates of growth of certain classes of monotone sequences. For example, which "grows quicker" as n increases, ln(n) or n? Our approach will be two-pronged. In the first stage, we will do some numerical background analysis to see what seems to be true for some values of n, usually "small" values of n. Then, after making a conjecture, we will do some general analysis to establish our conjecture for all possible values of n. Also, since it will be true that all the classes of sequences we look at will diverge to Infinity as n goes to Infinity, we will compare relative sizes of pairs of sequences. So we will be looking at pairs of sequences each time. The classes of sequences we investigate are:

1. logarithmic.......................$\ln(kn)$, $k = 1, 2, 3...$

2. polynomial ....$p(n) = n, n^2, n^3, ...$

3. exponential ....$e(n) = 2^n, 3^n, ...$, or, in general, a^n where $a > 1$

4. factorial ....$fac(n) = n!$

5. super - exponential ....$s(n) = n^n$

6. quadratic exponential ....$q(n) = 2^{n^2}$

Before getting to the analysis of each of these classes of sequences, let's consider the question of how one could go about comparing the rates of growth of two different sequences of numbers, say a_n and b_n. One common way to do this is to try to measure the rate of growth of the quotient of the two sequences term by term. (i.e. to measure the size of $\frac{a_n}{b_n}$ as n becomes "large".) Thus, we will need to evaluate $\lim\limits_{n \to \infty} \frac{a_n}{b_n}$.

DEFINITION . Let $\{a_n\}$ and $\{b_n\}$ be positive sequences.

$$\text{If } \lim_{n \to \infty} \frac{a_n}{b_n} = \begin{cases} L \neq 0, & \text{then we say that the rate of growth of } a_n \text{ is } L \text{ times that of } b_n \text{ ;} \\ \\ 0, & \text{then we say that the rate of growth of } b_n \text{ is much faster than that of } a_n \text{ ;} \\ \\ \infty, & \text{then we say that the rate of growth of } a_n \text{ is much faster than that of } b_n . \end{cases}$$

Notice, that if a_n is the sequence $a_n = 2n$ and b_n is the sequence $b_n = n$, then $\lim_{n \to \infty} \dfrac{a_n}{b_n} = 2$

which is what we expect since a_n is twice as large as b_n.

◼ NEW *MATHEMATICA* COMMANDS

There are no new *Mathematica* commands used in this lab. However, to refresh your memory, here is a list of the most frequently used commands for this lab.

tbl=Table[{b[n],a[n]},{n,n1,n2}]//N creates the table **tbl** and converts it to decimal form.
ListPlot[tbl,AspectRatio->1] plots the table **tbl**.
Limit[a[n]/b[n],n->Infinity] does just what you think it does.

◼ LOGARITHMIC GROWTH VERSUS POLYNOMIAL GROWTH

EXAMPLE 1. Let $a_n = \ln(n)$ and $b_n = n$.

Define a_n and b_n by **Enter**ing the commands:

```
Clear[a,b,tbl]
a[n_]:=Log[n]
b[n_]:=n
```
Now, let's compare the sizes of each of these sequences separately.

❏ **1.1.** Sketch the graph of the pairs of numbers (b[n],a[n]) for n=100 to 200.
HINT: Obtain a table of values of the pairs (b[n],a[n]) by **Enter**ing the command
tbl=Table[{b[n],a[n]},{n,100,200}]//N and then plot the graph by **Enter**ing the
command **ListPlot[tbl,AspectRatio->1]** (Recall that the x-axis is b[n]=n and the y-axis is
a[n]=Log[n].)

❏ **1.2.** In the above, as n changes from 100 to 200, how much does a[n] increase? How much does b[n] increase?

❏ **1.3.** Re-execute the above commands for the following range of values for n. In each case, state the growth of a[n] and b[n].
   (a) 500 - 600

   (b) 1000 - 1100

   (c) 5000 - 5100

❏ **1.4.** Based on your results so far, what can you say about the rate of growth of a[n] compared to b[n]? Justify your conclusion.

❏ **1.5.** What can you say about the rate of growth of a[n] relative to b[n]? Are they comparable, is a[n] much slower than b[n], or is a[n] much faster than b[n]? What can you say about the rate of growth of b[n] relative to a[n]? Did your answer to 1.4 "carry over" to large values of n?
   HINT. Find **Limit[a[n]/b[n],n->Infinity]** and **Limit[b[n]/a[n],n->Infinity]**

❏ **1.6.** Based on what you have seen, how does logarithmic growth compare to polynomial growth?

■ POLYNOMIAL GROWTH VERSUS EXPONENTIAL GROWTH

> **EXAMPLE 2.** Let $a_n = n^5$ and $b_n = 2^n$.

We first define the sequences by **Enter**ing the following commands:
Clear[a,b,tbl]
a[n_]:=n^5
b[n_]:=2^n

101

Now, let's examine the ratios $\dfrac{a[n]}{b[n]}$. To do so, we will create a table of these ratios and then plot them with respect to n. We create the table **tbl** by **Entering** the command

tbl = Table[a[n] / b[n], {n, 1, 50}] / /N and then plot these ratios, with respect to n, by **Entering** the command **ListPlot[tbl, AspectRatio- > 1]**   (The resulting plot is given below.)

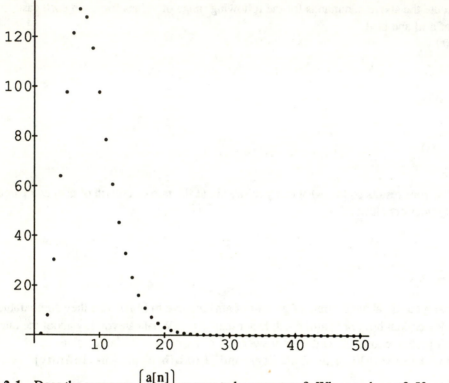

❏ **2.1.** Does the sequence $\left\{\dfrac{a[n]}{b[n]}\right\}$ appear to be monotone? Why or why not? If so, is it increasing

or is it decreasing? If not, is it possible for the sequence to have a limit? Why or why not?

❏ **2.2.** Based on this graph of the ratios, what does the limit (as n->+∞) of the ratios **appear** to be?

102

❑ **2.3.** Note that the graph has a "hump" in it. How can you be assured that the graph doesn't "jump up" again? Does it? Justify your answer.

HINT. Define f[x_]:=x^5/2^x and find where its derivative is 0 (<0).

❑ **2.4.** Is the sequence $\left\{\dfrac{a[n]}{b[n]}\right\}$ decreasing on the interval $(N,+\infty)$ for some N? If so, state the smallest such value of N.

❑ **2.5.** Re-create table **tbl** above for the following range of values for n.

   a) 100 - 200

   b) 500 - 600

   c) 1000 - 1100

How large does n have to be before $\dfrac{a[n]}{b[n]} < 1$? Don't worry about finding the best choice of n,

just find one choice.

❑ **2.6.** By taking an appropriate limit, compare the rate of growth of a[n] to that of b[n]. Justify your answer.

❑ **2.7.** Based on what you have seen, compare polynomial growth to exponential growth.

▪ EXPONENTIAL GROWTH VERSUS FACTORIAL GROWTH

EXAMPLE 3. Let $a_n = 5^n$ and $b_n = n!$.

Now, let's examine the ratios $\dfrac{a_n}{b_n}$. First, carefully define the sequences and then create a table

tbl of values of the ratios for values of n ranging from 1 to 30. Do a plot if you wish.

❑ **3.1.** Does the sequence $\left\{\dfrac{a_n}{b_n}\right\}$ appear to be monotone? Why or why not? If so, is it increasing

or is it decreasing?

❏ **3.2.** Based on this graph of the ratios, what does the limit (as n->+∞) of the ratios **appear** to be?

❏ **3.3.** Does it appear that $\left\{\dfrac{a_n}{b_n}\right\}$ is eventually monotone decreasing? Why or why not?

❏ **3.4.** How large would n have to be in order that $\left\{\dfrac{a_n}{b_n}\right\}$ is less than

(a) .5?

(b) .01?

(c) .000001?

❏ **3.5.** By taking an appropriate limit, compare the rate of growth of a_n to that of b_n. Justify your answer.

❏ **3.6.** Based on what you have seen, compare exponential growth to factorial growth.

▪ FACTORIAL GROWTH VERSUS SUPER-EXPONENTIAL GROWTH

EXAMPLE 4. Let $a_n = n!$ and $b_n = n^n$.

Now, let's examine the ratios $\dfrac{a_n}{b_n}$. First, carefully define the sequences and then create a table

tbl of values of the ratios for values of n ranging from 1 to 20. Do a plot if you wish.

❏ **4.1.** Does the sequence $\left\{\dfrac{a_n}{b_n}\right\}$ appear to be monotone? Why or why not? If so, is it increasing or is it decreasing?

❏ **4.2.** Based on this graph of the ratios, what does the limit (as n->+∞) of the ratios **appear** to be?

❏ **4.3.** Does it appear that $\left\{\dfrac{a_n}{b_n}\right\}$ is eventually monotone decreasing? Why or why not?

❏ **4.4.** How large would n have to be in order that $\left\{\dfrac{a_n}{b_n}\right\}$ is less than

(a) .5?

(b) .01?

(c) .000001?

❏ **4.5.** By taking an appropriate limit, compare the rate of growth of a_n to that of b_n. Justify your answer.

❏ **4.6.** Based on what you have seen, compare factorial growth to super-exponential growth.

◾ SUPER-EXPONENTIAL GROWTH VERSUS QUADRATIC EXPONENTIAL GROWTH

> **EXAMPLE 5.** Let $a_n = n^n$ and $b_n = 2^{n^2}$.

Now, let's examine the ratios $\dfrac{a_n}{b_n}$. First, carefully define the sequences and then create a table **tbl** of values of the ratios for values of n ranging from 1 to 20. Do a plot if you wish.

❏ **5.1.** Does the sequence $\left\{\dfrac{a_n}{b_n}\right\}$ appear to be monotone? Why or why not? If so, is it increasing or is it decreasing?

105

❏ **5.2.** Based on this graph of the ratios, what does the limit (as n->+∞) of the ratios **appear** to be?

❏ **5.3.** Does it appear that $\left\{\dfrac{a_n}{b_n}\right\}$ is eventually monotone decreasing? Why or why not?

❏ **5.4.** How large would n have to be in order that $\left\{\dfrac{a_n}{b_n}\right\}$ is less than

   (a) .5?

   (b) .01?

   (c) .000001?

❏ **5.5.** By taking an appropriate limit, compare the rate of growth of a_n to that of b_n. Justify your answer.

❏ **5.6.** Based on what you have seen, compare super-exponential growth to quadratic exponential growth.

▪ NOTE

The comparison of rates of growth of these types is **transitive**. That is, if polynomial growth is much slower than exponential growth and exponential growth is much slower than factorial growth, then polynomial growth is much slower than factorial growth. In general, if a_n's rate of growth is much slower than b_n's rate of growth and b_n's rate of growth is much slower than c_n's rate of growth, then a_n's rate of growth is much slower than c_n's rate of growth.

Series

NAME(S):   INSTRUCTOR:
CLASS: DATE:

■ INTRODUCTION

An **infinite series** is an expression having the form $\sum_{k=1}^{\infty} a_n = a_1 + a_2 + a_3 + \cdots + a_k + \cdots$. The

n – th **partial sum** of the series is $S_n = \sum_{k=1}^{n} a_k$. The series **converges** to the sum S if $\lim_{n \to \infty} S_n = S$.

If the limit S does not exist, then the series is said to **diverge**. In this lab, we will examine several series having non - negative terms (i.e. $a_n \geq 0$, n =1, 2, 3, 4,...)

■ NEW *MATHEMATICA* COMMANDS AND PROCEDURES

The only new procedure discussed in this lab is a way to define a function which "remembers" all of the function values it finds. (When you define a function by using f[x_]:= rhs, the value of the function is recomputed every time you ask for it.) This is particularly useful for defining recursive functions. An application of this is useful in examining the sum of a series. This will be illustrated in the examples. To define a function f which stores all values it finds, we use the form **f[x_]:=f[x]=rhs**

■ CONVERGENT SERIES

> **EXAMPLE 1.** Let $f(x) = \dfrac{x^2}{1 + e^{x-m}}$, where m = 1000.

❏ **1.1.** Carefully define m and f and then plot f(x) on the interval [0,1200]. State the command(s) you use as well as your results.

❏ **1.2.** Compute $\lim\limits_{x \to \infty} f(x)$. State the commands you **Enter** as well as the results you obtain.

Define the $n-$th partial sum of the series $\sum\limits_{k=1}^{\infty} a_k$, where $a_k = f(k)$, by **Entering** the following code:

Clear[sum]
sum[1]=f[1];
sum[n_]:=sum[n]=sum[n-1]+f[n]

❏ **1.3.** Create a table **tbl** containing the first 50 partial sums. State the commands you use as well as the results of **Enter**ing these commands.

❏ **1.4.** Using **ListPlot**, plot these sums with respect to n. What does this plot seem to suggest about the behavior of the series?

❏ **1.5.** Show that there is a number A such that if $x > A$, $f(x)$ is a decreasing function.

❏ **1.6.** Use the limit comparison test or the basic comparison test to determine whether this series converges or diverges.

❏ **1.7.** Show that the improper integral $\int_{1}^{\infty} f(x)\,dx$ is convergent and compute its value.

❏ **1.8.** Find the exact value of the smallest number N such that if n > N, then f(n) < .5.

❏ **1.9.** Comment on the following statement. "A positive termed series diverges if the first 10,000 terms become very large, otherwise it converges." Give supportive evidence for your position such as theorems presented in class or examples that we've examined.

▪ THE INTEGRAL TEST

Let $f(x)$ be a continuous, positive decreasing function with domain $[1,+\infty)$ and consider the series

$\sum_{n=1}^{\infty} a_n$, where $a_n = f(n)$ for all values of n. Then $\sum_{n=1}^{\infty} a_n$ converges if and only if $\int_{1}^{\infty} f(x)dx$ converges.

▪ THE HARMONIC SERIES

EXAMPLE 2. Consider the harmonic series $\sum_{k=1}^{\infty} \frac{1}{k}$.

Define the sequence **eulergamma[n]** $= 1 + \frac{1}{2} + \ldots \frac{1}{n} - \ln(n) = \sum_{k=1}^{n} \frac{1}{k} - \ln(n)$ by carefully **Entering** the following commands.

eulergamma[n_]:=(Sum[1/i,{i,n}] - Log[n]) // N

❑ **2.1.** Explain why **eulergamma[n]** is positive for all values of n.

❑ **2.2.** Show that **eulergamma[n+1]** is less than **eulergamma[n]** for all values of n.

110

Since $\{\mathbf{eulergamma[n]}\}_{n=1}^{+\infty}$ is decreasing and bounded below by 0, it must converge.

The actual value of $\underset{n\to\infty}{\text{Lim}}\ \mathbf{eulergamma[n]} = \underset{n\to\infty}{\text{Lim}}\left(\sum_{k=1}^{n}\frac{1}{k} - \ln(n)\right)$ is called Euler's constant

and is usually denoted by γ.

❏ **2.3.** The value of γ to three decimal places is .577. For what value of n does **eulergamma[n]** agree with the actual value of γ to three decimal places?

Notice that the n-th partial sum of the harmonic series, **har[n]**, satisfies **har[n]=eulergamma[n]+Log[n]**. Therefore, for large values of n,

$$\sum_{k=1}^{n}\frac{1}{k} = \mathbf{har[n]} \approx \gamma + Ln(n).$$

We can use this to determine how many terms are needed for $\sum_{k=1}^{N}\frac{1}{k}$ to attain the value M.

❏ **2.4.** Show that N must be greater than 600 for $\sum_{k=1}^{N}\frac{1}{k}$ to attain 7.

❏ **2.5.** For what value of N is $\sum_{k=1}^{N}\frac{1}{k} > 10$?

111

❑ **2.6.** For what value of N is $\displaystyle\sum_{k=1}^{N}\frac{1}{k} > 100$?

❑ **2.7.** Use this information to determine if the harmonic series converges or diverges.

▪ GEOMETRIC SERIES

A **geometric series** is a series of the form $\displaystyle\sum_{n=1}^{\infty} ar^{n-1}$. If $|r| < 1$, then the series converges to $\dfrac{a}{1-r}$.

If $|r| \geq 1$, the series diverges.

> **EXAMPLE 3.** Consider the series $\displaystyle\sum_{n=0}^{\infty}\frac{15^{n+4}}{20^{n-2}}$.

❑ **3.1.** Identify **a** and **r**. Does the series converge? If so, find its limit.

> **EXAMPLE 4.** Approximate all values of x for which $\displaystyle\sum_{n=0}^{\infty} (x^2 + 6x + 7)^n$ converges.

❑ **4.1.** Identify **a** and **r**.

❑ **4.2.** Approximate the values of x for which $|r| < 1$ by graphing r (as a function of x), or $|r|$, and using **FindRoot**.

Taylor Polynomials

NAME(S):   **INSTRUCTOR:**
CLASS: **DATE:**

▪ INTRODUCTION

Let $f(x)$ be a function which is infinitely differentiable at $x = a$. Then, the **Taylor Series** expansion

of $f(x)$ about $x = a$ is $f(x) = f(a) + f'(a)(x - a) + \dfrac{f''(a)}{2}(x - a)^2 + \cdots + \dfrac{f^{(n)}(a)}{n!}(x - a)^n + \cdots$. The

Taylor Polynomial of degree n is $\mathbf{taylorpolynf}(x) = f(a) + f'(a)(x - a) + \dfrac{f''(a)}{2}(x - a)^2 + \cdots$

$+ \dfrac{f^{(n)}(a)}{n!}(x - a)^n$. (Actually, if $f^{(n)}(a) = 0$, then the degree of $taylorpolynf(x)$ is less than n.)

Note that $taylorpolynf(x)$ approximates $f(x)$.

Many situations arise in mathematics which involve approximating certain functions by other functions which are simpler and perhaps "nicer" in some respects. In this lab, we will examine the approximation of functions differentiable at x=a by Taylor Polynomials. Evaluations of polynomials are easy and straightforward since the only operations required are addition and multiplication. The evaluation for certain other functions is not so easy.

NOTE. If a=0, then the series is a **Maclaurin Series** and the polynomial is a **Maclaurin Polynomial.**

In this lab session we will be concerned with generating the n-th degree Taylor Polynomial for a given function $f(x)$ and then investigate to find the largest interval of x values for which the Taylor Polynomial will fit within a given tolerance of $f(x)$. Since we know that the Taylor Polynomial agrees with $f(x)$ at the expansion point, this interval will contain the point of expansion in its interior. In more formal terms, we want to answer the following:

Given: 1) a function $f(x)$
        2) a point $x = a$ ....the point of expansion
        3) a value n ....the degree of the Taylor Polynomial
        4) error ....the value of the error term

Find: 1) the n-th degree Taylor Polynomial, $taylorpolynf(x)$, of $f(x)$ about the point $x = a$
       2) the largest interval containing the point $x = a$ for which $|f(x) - taylorpolynf(x)| \leq error$.

▣ NEW *MATHEMATICA* COMMANDS

1. **Series[f,{x,x_0,n}]** generates a power series expansion for f about the value x = x_0 up to order n.

2. **Normal[expr]** converts **expr** to a normal expression, from a variety of special forms. In particular, if expr is a series, Normal[expr] removes the "big O" term of the series and returns a polynomial.

3. If y= f(x) is a function that has a Taylor series expansion about x=a, then **taylorseriesf(x)** will denote the Taylor Series for f(x) and **taylorpolynf(x)** will denote the n-th degree Taylor Polynomial for f(x). *Mathematica* can be used to obtain **taylorpolynf(x)** for any reasonable value of n.

If f(x) has been defined, then the general command to create taylorpolynf(x) is
> **taylorpolynf[x_]=Normal[Series[f[x],{x,a,n}]]**

where the values of **n** and **a** are inserted before **Entering** the command.
NOTE. The point **a** is the point of expansion and **n** is the desired power or degree.

For example, if f(x) has been defined, then **taylorpoly5f[x_]=Normal[Series[f[x],{x,0,5}]]** yields the 5-th Taylor Polynomial for f(x) about x=0 (i.e. the 5-th **Maclaurin Polynomial** for f(x)), while **taylorpoly9f[x_]=Normal[Series[f[x],{x,.5,9}]]** yields the 9-th Taylor Polynomial of f(x) about x=.5.

▣ TAYLOR POLYNOMIALS

EXAMPLE 1. $f(x) = \sqrt{x^2+1}$ and a = 0.

❑ **1.1.** Find the Taylor polynomial of degree 3 of f(x) expanded about x = a by **Entering** the following commands:
Clear[f,taylorpoly3f]
f[x_]:=Sqrt[x^2+1]
taylorpoly3f[x_]=Normal[Series[f[x],{x,0,3}]]

❑ **1.2.** Plot y = f(x), y = taylorpoly3f(x), and y=1.001 simultaneously on the interval [-3,3]. (Use different GrayLevels to distinguish the graphs.)

114

❑ **1.3.** Find taylorpoly10f(x) and then plot y = f(x) , y = taylorpoly3f(x) and y = taylorpoly10f(x) simultaneously on the interval [-1,1]. (Again, use different GrayLevels to distinguish the graphs.) Expand the plot-interval width to [-3,3] and then to [-10,10]. Note the difference between taylorpoly3f(x) and taylorpoly10f(x). NOTE. You may want to restrict the range of y by using the Plot option **PlotRange -> {a,b}**

❑ **1.4.** Complete the entries in the following table.

n	error	expansion point	maximum interval for which $\|f(x) - taylorpolynf(x)\| \leq$ error	interval width
2	.001	0	[-.302423 , .302423]	.604846
3	.001	0		
4	.001	0		
10	.001	0		
20	.001	0		

HINT. Define an error function, errorn(x) = f(x) - taylorpolynf(x). Then simultaneously plot y=errorn(x), y=-.001, and y = .001 on the same interval. You may need to adjust the plot interval in order to get a good approximation of the x-value where the curve y=errorn(x) and the line y = .001 (and/or y=-.001) intersect. Then use the **FindRoot** command to solve the equation errorn(x) = .001 (and/or errorn(x) = -.001). For n=2, we used the following commands.

```
Clear[taylorpoly2f,error2]
taylorpoly2f[x_]=Normal[Series[f[x],{x,0,2}]]
error2[x_]=f[x] - taylorpoly2f[x]
Plot[{error2[x],-.001,.001},{x,-1,1}]
FindRoot[error2[x]==-.001,{x,-.3}]
FindRoot[error2[x]==-.001,{x,.3}]
```

EXAMPLE 2. $g(x) = e^{-x^2}$, a=1.5.

❑ **2.1.** Find the Taylor Polynomial of degree 10 of g(x) expanded about x = a.

❑ **2.2.** Plot y = g(x) and y = taylorpoly10g(x) simultaneously on the interval [-.5,3.5]. Expand the plot-interval width as much as appears feasible.

❑ **2.3.** Find taylorpoly20g(x) and then plot y = g(x) , y = taylorpoly10g(x) and y = taylorpoly20g(x) simultaneously on the interval [-1.5,4.5].   (Again, use different GrayLevels to distinguish the graphs and restrict the range, if necessary.) Expand the plot-interval width as much as appears feasible and note the difference between taylorpoly10g(x) and taylorpoly20g(x).

❑ **2.4.** Complete the entries in the following table.

n	error	expansion point	maximum interval for which lg(x) - taylorpolyng(x) l ≤ error	interval width
2	.001	1.5	[1.30809,1.68165]	.37356
3	.001	1.5		
10	.001	1.5		
20	.001	1.5		
30	.001	1.5		

Approximations by Taylor Polynomials

NAME(S):   INSTRUCTOR:
CLASS: DATE:

■ INTRODUCTION

Let $f(x)$ be a function which is differentiable $(n + 1)$ times on an open interval I and let $a \in I$. Then, for each $x \in I$, there exists a number w between x and a such that

$$f(x) = f(a) + f'(a)(x - a) + \frac{f''(a)}{2}(x - a)^2 + \cdots + \frac{f^{(n)}(a)}{n!}(x - a)^n + \frac{f^{(n+1)}(w)}{(n + 1)!}(x - a)^{n+1} . \text{ We write}$$

$$f(x) = \text{taylorpolynf}(x) + \text{rnf}(x), \text{ where } \textbf{taylorpolynf}(x) = f(a) + f'(a)(x - a) + \frac{f''(a)}{2}(x - a)^2 + \cdots$$

$$+ \frac{f^{(n)}(a)}{n!}(x - a)^n \text{ and } \textbf{rnf}(x) = \frac{f^{(n+1)}(w)}{(n + 1)!}(x - a)^{n+1} . \text{ The Taylor Polynomial of degree n (or less)}$$

is **taylorpolynf(x)** and the remainder is **rnf(x)**. Note that taylorpolynf(x) approximates f(x) and the error in that approximation is rnf(x).

Many situations arise in mathematics which involve approximating certain functions by other functions which are simpler and perhaps "nicer" in some respects. In this lab, we will examine the approximation of functions differentiable at x=a by Taylor Polynomials. The approximating Taylor Polynomials will be best **near x=a**. Also, it is generally true that the larger the value of n, the better the approximation is. Evaluations of polynomials are easy and straightforward since the only operations required are addition and multiplication. The evaluation for certain other functions is not so easy.

In this lab, we will examine two types of problems that typically arise when approximating a function f(x) by a Taylor Polynomial taylorpolynf(x) of degree n (or less).

1. Given: f(x), a, and n.
   Find: the error that results when taylorpolynf(x) is used to approximate f(x) **near x=a**.
2. Given: f(x), a , and a value, which we will call error, that specifies the degree of accuracy.
   Find: how large n must be in order for | taylorpolynf(x) - f(x) | < error **near x=a**.

NOTE. If a=0, then the series is a **Maclaurin Series** and the polynomial is a **Maclaurin Polynomial**.

117

■ NEW *MATHEMATICA* COMMANDS

As you noted in Lab # 22, if y= f(x) is a function that has a Taylor series expansion about x=a, then **taylorseriesf(x)** will denote the Taylor series for f(x) and **taylorpolynf(x)** will denote the n-th degree Taylor polynomial for f(x). *Mathematica* can be used to obtain **taylorpolynf(x)** for any reasonable value of n. If f(x) has been defined, then the general command to create taylorpolynf is

taylorpolynf[x_]=Normal[Series[f[x],{x,a,n}]]

where the values of **n** and **a** are inserted before **Entering** the command.
NOTE. The point **a** is the point of expansion and **n** is the desired power or degree.

For example, if f(x)=ln(1+x), then **taylorpoly5f[x_]=Normal[Series[f[x],{x,0,5}]]** yields the 5-th Taylor Polynomial for f(x) about the point x=0. Similarly, **taylorpoly9f[x_]=Normal[Series[f[x],{x,.5,9}]]** yields the 9-th Taylor Polynomial of f(x) about x=.5.

As you recall, the command **D[f[x],{x,n}]** denotes $f^{(n)}(x)$, the $n-th$ derivative of f(x).

■ APPROXIMATIONS BY TAYLOR POLYNOMIALS

EXAMPLE 1.   f(x)=ln(1+x)

Enter each of the following commands.
Clear[f,taylorpoly1f,taylorpoly2f,taylorpoly3f,taylorpoly4f]
f[x_]:=Log[1+x]
Enter each of the following commands and indicate the result obtained.

❏ **1.1.   taylorpoly1f[x_]=Normal[Series[f[x],{x,0,1}]   ]**

❏ **1.2.   taylorpoly2f[x_]=Normal[Series[f[x],{x,0,2}]   ]**

❏ **1.3.   taylorpoly3f[x_]=Normal[Series[f[x],{x,0,3}]   ]**

❏ **1.4.   taylorpoly4f[x_]=Normal[Series[f[x],{x,0,4}]   ]**

118

❏ **1.5. Plot[{f[x], taylorpoly1f[x], taylorpoly2f[x], taylorpoly3f[x], taylorpoly4f[x]}, {x,-.9,7}, PlotStyle->{GrayLevel[0.], GrayLevel[.2], GrayLevel[.4], GrayLevel[.6], GrayLevel[.8]}]**

NOTE. Instead of the above **PlotStyle** option, you can use
PlotStyle->Map[GrayLevel,Range[0.,.8,.2]]]

❏ **1.6.** As in 1.5. above, but change the interval to [-.9,.7]

❏ **1.7.** As in 1.5. above, but change the interval to [-.2,.2]

❏ **1.8.** As in 1.5. above, but change the interval to [.6,.8]

❏ **1.9.** What is(are) the "moral(s) of the story" in 1.5, 1.6, 1.7 and 1.8?

EXAMPLE 2. If $f(x)=\ln(1+x)$, then find the error that results when taylorpolynf(x) is used to approximate f(x) if taylorpolynf(x) is expanded about a=0.

Consider the following question: What is a reasonable upper bound on the error that results when approximating ln(1.7) by taylorpoly9f(x)? Note that f(0.7)=ln(1.7) and thus we want to approximate f(0.7) by taylorpoly9f(0.7) and bound the resulting error.

Recall: $\left| f(x) - \text{taylorpolynf}(x) \right| = \left| \dfrac{f^{(n+1)}(w)(x-a)^{n+1}}{(n+1)!} \right|$ for some w between x and a,

where $a = 0$. In this example, $x = .7$ and $n = 9$. Hence, we need to find a bound for $\left| f^{(10)}(w) \right|$,

where w is between 0 and .7. Recall that the *Mathematica* function **Log** represents the natural log function.

❏ **2.1.** Find the maximum value of $\left| D[\ f[x],\{x,10\}\ ] \right|$ on the interval [0,0.7].
HINT. After defining f, let **df10=D[f[x],{x,10}]** and then plot **df10** on [0,0.7].

❏ **2.2.** Find a reasonable upper bound for $\left| f(0.7) - \text{taylorpoly9f}(0.7) \right|$. Justify your answer.

❏ **2.3.** Find the value of taylorpoly9f(x) at x=0.7.

❏ **2.4.** Graph f(x) and taylorpoly9f(x) on [0,1.4] using different GrayLevels to distinguish the graphs.

❑ **2.5.** *Mathematica* can approximate ln(1.7) to any desired degree of accuracy. Compare *Mathematica*'s value for |f(0.7) - taylorpoly9f(0.7)| to your estimate for |f(0.7) - taylorpoly9f(0.7)| .

❑ **2.6.** Suppose we change n to 15. Find a reasonable upper bound on the error that results from approximating ln(1.7) by evaluating taylorpoly15f(x) at x=0.7.
HINT. Change taylorpoly9f(x) to taylorpoly15f(x) and do the same type of analysis.

❑ **2.7.** Graph f(x), taylorpoly9f(x), and t15f(x) on [.695,.705] using different GrayLevels to distinguish the graphs.

EXAMPLE 3. If f(x)=ln(1+x), then how large must n be if we want taylorpolynf(x) to approximate f(0.7) to within 5 place accuracy? (i.e. How large must n be in order to assure that |f(0.7) - taylorpolynf(0.7)| < accuracy, where accuracy=.000005?) Assume taylorpolynf(x) is expanded about a=0.

By the Remainder Theorem, error $= \left| \dfrac{f^{(n+1)}(w)(x-a)^{n+1}}{(n+1)!} \right|$ for some w between x and a.

Now $x = .7$ and $a = 0$, so error $= \left| \dfrac{f^{(n+1)}(w)(.7)^{n+1}}{(n+1)!} \right|$ for some w between 0 and 0.7 .

121

Thus, if we could find an upper bound U for $\left| f^{(n+1)}(x) \right|$ for x between 0 and 0.7, then we

could substitute this value for $f^{(n+1)}(w)$ and then have the inequality error $\leq \left| \dfrac{U\ (.7)^{n+1}}{(n+1)!} \right|$

which is an expression containing only n. Note that the maximum error is $\left| \dfrac{U\ (.7)^{n+1}}{(n+1)!} \right|$.

❏ **3.1.** Find a general formula for $\left|\ D[f[x],\{x,n\}]\ \right|$.
HINT. f(x) is already defined, so **Enter** the command **dfn=D[f[x],{x,n}]** for n=1,2,3,... and watch for the pattern.

dfn=

$\left|\ \textbf{dfn}\ \right| =$

❏ **3.2.** Find an upper bound U for $\left|\text{dfn}\right|$ on the interval [0,.7] that does not depend upon the variable x. At what value of x does this occur? NOTE: We are lucky in the case of f(x)=ln(1+x) since the maximum of $\left|f(x)\right|$ occurs at one of the endpoints of [0,.7]. (In general, we would have to use techniques developed in the first calculus course to find an upper bound.)

U =

occurs at x=

❏ **3.3.** Express the maximum error as a function of n. We can do this because of what we have done in **3.1** and **3.2**. Justify your answer.

maxerror[n]=

❏ **3.4.** Find the smallest value of n for which maxerror < accuracy. (i.e. Find the smallest n for which (maxerror - accuracy) < 0.
HINT. Define maxerror(n) and then generate a table of values of (maxerror(n) - accuracy) by **Enter**ing the command **tbl=Table[maxerror[n]-accuracy,{n,1,20}]**

smallest n =

❏ **3.5.** How large must n be if we want taylorpolynf to approximate f(0.7) to within 8 place accuracy? (i.e. How large must n be in order to assure that l f(0.7) - taylorpolynf(0.7) l < accuracy, where accuracy = .000000005?) HINT. Change the value of **accuracy** and do the same type of analysis that we did for 5 place accuracy.

122

Applications of Power Series

NAME(S):   **INSTRUCTOR:**
CLASS: **DATE:**

▣ INTRODUCTION

The purpose of this lesson is to examine two applications of series. The first application is to illustrate how to obtain series solutions to several differential equations. The second application is to illustrate how series may be used to obtain rational approximations of certain functions.

▣ NEW *MATHEMATICA* COMMANDS

Recall that **Series[f,{x,x_0,n}]** generates a power series expansion for f about the value $x = x_0$ up to order n and **Normal[expr]** converts **expr** to a normal expression, from a variety of forms. In particular, if expr is a series, then Normal[expr] removes the "big O" term of the series and returns a polynomial.

If two power series **ser1** and **ser2** are equal, then their corresponding coefficients must be the same. **LogicalExpand[ser1==ser2]** equates the coefficients of the equal series **ser1** and **ser2**. The result is a system of equations.

▣ SERIES SOLUTIONS TO DIFFERENTIAL EQUATIONS

> **EXAMPLE 1.** Find a function y(x) that satisfies the ordinary differential equation
> y"(x)-4y'(x)-5y(x)=0 and the conditions y(0)=1 and y'(0)=5.

❏ **1.1.** Let y(x) denote the solution to the problem. Use the **Series** command to find the first eight terms of the power series for y(x) about x=0 and name the result **ser**.

❏ **1.2.** Since y(0)=1 and y'(0)=5, replace y[0] by 1 and y'[0] by 5 in **ser** and name the result **sery** by **Enter**ing the command   sery = ser /. { y[0]->1 , y'[0]->5 }

Notice that **sery** must satisfy the differential equation y"(x)-4y'(x)-5y(x)=0 so we obtain the equation:

$$\frac{d^2}{dx^2}\,\mathbf{sery} - 4\,\frac{d}{dx}\,\mathbf{sery} - 5\,\mathbf{sery} = 0$$

❏ **1.3.** Proceeding with *Mathematica*, define **equation** to be the above equation by **Enter**ing **equation=D[sery,{x,2}]-4 D[sery,x]-5 sery==0**

❏ **1.4.** Identify the unknowns in **1.3.**

❏ **1.5.** We can solve for the unknowns by equating coefficients and then solving the resulting system of (linear equations). One way of accomplishing this is to first **Enter** the command **equals=LogicalExpand[equation]**

then solve the resulting equations by **Enter**ing **roots=Solve[equals]**

and finally substitute these values into **sery** by **Enter**ing

solution=sery /. roots[[1]]

❑ **1.6.** Compare **solution** to the series for $y = e^{5x}$ about $x = 0$.

❑ **1.7.** Verify that $y = e^{5x}$ is a solution to the differential equation $y''(x) - 4y'(x) - 5y(x) = 0$ that satisfies the required initial conditions.

▣ APPROXIMATION BY RATIONAL FUNCTIONS

> **EXAMPLE 2.** Let $f(x) = \dfrac{1}{1 - \sin(x)}$.

The purpose of this example is to find the best rational approximation of the form

$$g(x) = \frac{b[0] + b[1]x + b[2]x^2}{1 + b[3]x} \quad \text{for the function } f(x) = \frac{1}{1 - \sin(x)} \quad \text{on an open interval containing}$$

$x = 0$.

This is equivalent to finding b[0], b[1], b[2], and b[3] so that f(x) and g(x) agree as closely as possible on a neighborhood containing zero.

We proceed with *Mathematica*. Begin by defining **f** and **expression**:

Clear[f,g,h,b,expression]
f[x_]=1/(1-Sin[x])
expression=(b[0]+b[1] x+b[2] x^2)/(1+b[3]x)

Notice that we need to find four numbers b[0], b[1], b[2], and b[3] to solve the problem.

❑ **2.1.** Use *Mathematica* to obtain the first four terms of the Taylor series for **f[x]** and **expression** by carefully **Entering** each command and indicating the result.
Clear[ser1,ser2]
ser1=Series[f[x],{x,0,3}]

ser2=Series[expression,{x,0,3}]

❏ **2.2.** To obtain the best approximation, **ser1** and **ser2** should be as much alike as possible. Use **LogicalExpand** to equate the coefficients of **ser1** and **ser2** by carefully **Enter**ing:
Clear[equation]
equations=LogicalExpand[ser1==ser2]

Notice that the system of four equations in four unknowns may be solved by hand for the variables b[0], b[1], b[2], and b[3]. However, it is easier to use *Mathematica*'s **Solve** command.

❏ **2.3.** Use **Solve** to solve the system of equations **equations** by carefully **Enter**ing:
Clear[roots]
roots=Solve[equations]

❏ **2.4.** To obtain g(x), replace b[0], b[1], b[2], and b[3] by the solutions to **equation**:
g[x_]=Simplify[expression/.roots[[1]]]

Let's compare the rational approximation g(x), the Taylor polynomial approximation **ser1**, and the actual function f(x).

❏ **2.5.** First, however, since **ser1** is a series, not a function, use *Mathematica* to create a function for **ser1** by **Entering**:
h[x_]=Normal[ser1]

❏ **2.6.** Compare the graph of the rational approximation g(x) to the graph of f(x) on the interval [-1,1]. Use different GrayLevels to distinguish the graphs.

❏ **2.7.** Compare the graph of the Taylor polynomial h(x) to the graph of f(x) on the interval [-1,1]. Use different GrayLevels to distinguish the graphs.

❏ **2.8.** Which of the two is the better approximation? Why?

❏ **2.9.** To compare the actual error, graph $|g(x)-f(x)|$ and $|h(x)-f(x)|$ on the interval [-1,1]. Use different GrayLevels to distinguish the graphs.

❏ **2.10.** Modify the following command to find a value of r such that the maximum error between g(x) and f(x) on the interval [-r,r] is at most .001

```
Plot[Abs[g[x]-f[x]],{x,-.5,.5},PlotRange->All]
```

EXAMPLE 3. The differential equation $y^{(4)} + \lambda y'' = 0$ models the buckling of a long column (like the bending of a beam).

Use power series to determine the solutions using the indicated boundary conditions. In each case, plot the resulting solution.

❏ **3.1.** y(0)=0,y'(0)=0 (Fixed end at 0), y''(1)=0,y'''(1)=0 (Free end at 1)

❏ **3.2.** y(0)=0,y''(0)=0 (Simple support at 0), y'(1)=0, y'''(1)=0 (Sliding clamped end)

Parametric Equations and Quadratic Equations

NAME(S):   INSTRUCTOR:
CLASS: DATE:

▣ INTRODUCTION—PARAMETRIC EQUATIONS

Using parametric equations, one can build some very interesting and varied graphs. In this lab we will study the graphs of curves given by parametric equations. Some of the features we will try to identify for a particular curve are:

1. The direction of the tangent vector at various points along the curve, and
2. The points for which the curve has a vertical tangent line.

Let's examine each of these features separately. To determine the tangent line to a curve, we must be able to find its slope. Suppose that x and y have been defined as functions of t on some interval. Then, we know that $\dfrac{dy}{dx} = \dfrac{y'(t)}{x'(t)}$. Thus, the equation of the tangent line for $t = t_0$ can be written as

$l(t) = \{x(t_0), y(t_0)\} + t\{x'(t_0), y'(t_0)\}$. Also, because of the above form of $\dfrac{dy}{dx}$, the vertical tangent lines will occur when $x'(t) = 0$ and $y'(t) \neq 0$.

▣ NEW *MATHEMATICA* COMMANDS

There are two new *Mathematica* commands that we will use in this session. They are:

1. **ParametricPlot[{x[t],y[t]},{t,tstart,tfinish}]**
   This command is a standard *Mathematica* command and plots the points (x[t],y[t]) for t in the interval [tstart,tfinish]. Before the command is **Entered**, x[t], y[t], **tstart**, and **tfinish** have to be defined. To see an example, **Enter** the following commands .
   x[t_]:=Sin[t]
   y[t_]:=Cos[2t]
   ParametricPlot[{x[t],y[t]},{t,0,2Pi}]

2. **paramplot[{x,y},{tstart,tfinish},tset].**
   The command **paramplot** is not a standard *Mathematica* command. It has been defined for use in this lab and is available from your instructor. This command also plots the points (x[t],y[t]) for t in

the interval [tstart, tfinish]. In addition, for each set tset = $\{t_1, t_2, \ldots, t_n\}$ of t values, the command will plot the tangent vector at each point $\left(x(t_k), y(t_k)\right)$ for $k = 1, 2, \ldots n$. The tangent vector points in the direction of the curve. To see an example, **Enter** the following command (after x and y have been defined as above).

paramplot[{x, y}, {0, 2 Pi}, {1, 2, 3}]

■ PARAMETRIC EQUATIONS

EXAMPLE 1. $p(t) = \left(\sin\left(\sqrt{5} \, \sin(t)\right), \cos\left(\sqrt{2} \, t\right)\right)$

❑ **1.1.** Plot p(t) for t in [0, 6π].
HINT. After **Enter**ing the commands below, use **ParametricPlot**
Clear[x,y]
x[t_]:=Sin[Sqrt[5] Sin[t]]
y[t_]:=Cos[Sqrt[2] t]

❑ **1.2.** Describe your reaction to the graph.

EXAMPLE 2. $c(t) = \left(\sin(t), t^{1/3}\right)$

❑ **2.1.** Plot c(t) for t in the interval [-3π, 5π] and plot the tangent vectors for t = -5, -2, 1, and 7.
HINT. After **Clear**ing x and y, define x and y and then use **paramplot**.

130

❏ **2.2.** Locate the value(s) of t in the interval [-3π,5π], if any, for which the tangent line is vertical.

EXAMPLE 3. $c(t) = \left((t-1)(t-5)(t-9), t^{2/3}\cos(t) \right)$

❏ **3.1.** Plot c(t) for t in the interval [-1,10] and plot the tangent vectors when t = -1, 1, 6, and 9.

❏ **3.2.** Locate the value(s) of t in the interval [-1,10], if any, for which the tangent line is vertical.

EXAMPLE 4. $p(t) = \left(\sin(t) + \cos(3t), \cos^2(t) \right)$

❏ **4.1.** Plot p(t) for t in the interval [0, 2π] and plot the tangent vectors when t = -1, 1, 6, and 9.

❏ **4.2.** Locate the value(s) of t in the interval [0,2π], if any, for which the tangent line is vertical.

■ INTRODUCTION—QUADRATIC EQUATIONS

The general form of a quadratic equation in two variables is:

$$a x^2 + b x y + c y^2 + d x + e y + f = 0$$

By letting $p(x,y) = a x^2 + b x y + c y^2 + d x + e y + f$, the equation becomes $p(x,y) = 0$.

By a rotation of axes, the xy term can be forced to zero. A rotation of axes is accomplished by
1. replacing x with *x cosθ - y sinθ* and
2. replacing y with *x sinθ + y cosθ*

where θ is the angle of rotation. (**θ can always be chosen such that 0<θ<π/2.**) Thus, to perform the rotation, we need to determine sinθ and cosθ. These values can be found as follows:

$$\sin(\theta) = \sqrt{\frac{1 - \cos(2\theta)}{2}} \text{ and } \cos(\theta) = \sqrt{\frac{1 + \cos(2\theta)}{2}}$$

The value of cos(2θ) can be expressed in terms of the coefficients a, b and c as follows:

$$\cos(2\theta) = \frac{a-c}{hyp} * \text{Sign}(b) \text{ where } hyp = \sqrt{(a-c)^2 + b^2} \text{ and } 0 < \theta < \frac{\pi}{2}.$$

■ QUADRATIC EQUATIONS

> **EXAMPLE 5.** $16 x^2 - 24 x y + 9 y^2 - 90 x - 120 y = 0$

❏ **5.1. Enter** each of the following commands. Indicate each output and what it represents.
Clear[p,x,y,a,b,c,d,e,f]

p[x_,y_]:= a x^2 + b x y + c y^2 + d x + e y + f

a=16; b=-24; c=9; d=-90; e=-120; f=0

hyp=Sqrt[(a-c)^2 + b^2]

cos2theta=(a-c)*Sign[b]/hyp

sintheta=Sqrt[(1-cos2theta)/2]

costheta=Sqrt[(1+cos2theta)/2]

theta=ArcSin[sintheta] 180/Pi//N

xp=x costheta - y sintheta

yp=x sintheta + y costheta

p[xp,yp]

Chop[Simplify[p[xp,yp]]//N]

❏ **5.2.** Find the **angle of rotation** that eliminates the xy term.

❏ **5.3.** Find the **equivalent equation** for this curve that does not contain an xy term.

❏ **5.4.** Find the **class** to which the curve belongs. (i.e. circle, parabola,...) Justify your answer.

EXAMPLE 6. $3x^2 + 4xy - 4 = 0$

❏ **6.1.** Find the **angle of rotation** that eliminates the xy term.

❏ **6.2.** Find the **equivalent equation** for this curve that does not contain an xy term.

❑ **6.3.** Find the **class** to which the curve belongs. (i.e. circle, parabola,...) Justify your answer.

EXAMPLE 7. $x^2 + 2\sqrt{3}\,xy + 3y^2 + 2\sqrt{3}\,x - 2y = 0$

❑ **7.1.** Find the **angle of rotation** that eliminates the xy term.

❑ **7.2.** Find the **equivalent equation** for this curve that does not contain an xy term.

❑ **7.3.** Find the **class** to which the curve belongs. (i.e. circle, parabola,...) Justify your answer.

EXAMPLE 8. If a quadratic equation contains both an x^2 term and a y^2 term then it cannot be a parabola.

❑ **8.1.** Determine whether or not this statement is true. Justify your answer.

Graphing in Polar Coordinates

NAME(S):   **INSTRUCTOR:**
CLASS: **DATE:**

▣ INTRODUCTION

The rectangular coordinate system is convenient, and quite satisfactory, when graphing most plane curves. However, there are many plane curves that can best be described, and graphed, by using the **polar coordinate system.** Suppose the **pole** is denoted by the point O in the plane. If P is a point in the plane, then P is represented by a pair (r,θ), where r is the distance of P from O and θ is the angle formed (measured counterclockwise) between the **polar axis** and the ray OP. For many curves in this coordinate system, $r=r(\theta)$. An example is $r = 1+2 \cos(\theta)$. In this lab, we will examine curves in the polar coordinate system. NOTE. For simplicity we will use t rather than θ.

▣ NEW *MATHEMATICA* COMMANDS

PolarPlot is used to plot curves expressed in polar coordinates. The general format for this command is: **PolarPlot[f[t],{t,tmin,tmax}]** This plots the graph of r=f(t) for $tmin \le t \le tmax$. An example of this command is **PolarPlot[Sin[2 t],{t,0,2Pi}]** which plots the graph of r=sin(2t) for $0 \le t \le 2\pi$. This is illustrated in Example 1 below.

In order to use PolarPlot, we have to load the graphics package Graphics.m. We have the command to load this package in the initialization cell. On occasion when initializing, *Mathematica* will tell you *Mathematica* **can't find Graphics.m**. If that happens, you will need to help *Mathematica* find it. You simply need to open each of the following files in the order specified: *Mathematica* f, Packages, Graphics, and finally, Graphics.m. At that point, *Mathematica* will load Graphics.m, and you are "on your way".

Instead of using **PolarPlot**, you can use **ParametricPlot**. In converting from polar coordinates **(r,t)** to rectangular coordinates **(x,y)**, you use x=r **cos(t)** and y=r **sin(t)**. Thus, if r=f(t), instead of the command **PolarPlot[f[t],{t,tmin,tmax}]** you can use the command **ParametricPlot[{f[t] Cos[t],f[t] Sin[t]},{t,tmin,tmax}]** to plot the same graph. You probably will want to use the option **AspectRatio->Automatic** to avoid distortion.

■ **EXAMPLES**: r=m sin(nt) and r=kt

EXAMPLE 1. r = m sin(nt), $0 \leq t \leq 2\pi$.

To plot r = sin(2t) for $0 \leq t \leq 2\pi$, one **Enters** the command **PolarPlot[Sin[2 t],{t,0,2Pi}]**
Upon doing so, the following graph is obtained.

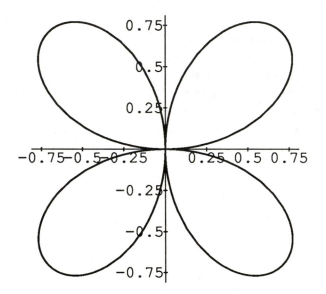

❑ **1.1.** What would be a good name for the graph?

❑ **1.2.** Plot r = sin(4t) for $0 \leq t \leq 2\pi$. What would be a good name for the graph?

❑ **1.3.** If the same pattern follows, what would be a good name for the graph of $r = \sin(5t)$ for $0 \leq t \leq 2\pi$. Does a plot verify your **conjecture**? If not, why not?

❑ **1.4.** Plot $r = 2 \sin(t)$ for $0 \leq t \leq 2\pi$. What would be a good name for the graph?

❑ **1.5.** How does the graph of $r = 4 \sin(t)$ for $0 \leq t \leq 2\pi$ compare to the graph in **1.4**?

❑ **1.6.** How does the graph of $r = 4 \sin(2t)$ for $0 \leq t \leq 2\pi$ compare to the graph preceding **1.1**?

❑ **1.7.** Based on the preceding questions (and their answers, of course), describe the graph of $r = m \sin(nt)$ for $0 \leq t \leq 2\pi$, where n is a positive integer. You may have to do another plot (or two) to satisfy yourself that your **conjecture** is correct.

❑ **1.8.** Would your answer to **1.7** change if n is a negative integer? If so, how?

❑ **1.9.** How would you **prove** your conjecture in **1.7** above? (We aren't asking you to actually prove it, but to state how you would go about proving it.)

EXAMPLE 2.   r=kt

❏ **2.0.** Plot $r = \dfrac{t}{2}$, $t \geq 0$. What is a good name for the graph? (This graph is associated with the name of **Archimedes**.)

❏ **2.1.** Plot $r=2t$, $t\geq 0$. How does this graph compare to the graph in 2.0 above?

❏ **2.2.** Describe the graph of $r=kt$, $t\geq 0$ for $k>0$. How does the value of k affect the graph?

❏ **2.3.** How does the graph of $r=kt$, $t\geq 0$ compare to the graph of $r=-kt$, $t\geq 0$ for a fixed $k\neq 0$?

■ LIMAÇONS

Limaçons are curves which are also best described with polar coordinates. A limaçon has equation $r = a + b \sin(t)$ or $r = a + b \cos(t)$, where $a\neq 0$. Since $r= a + b \cos(t)$ can be written as $r = a(1 + b/a \cos(t))$, we can envision **a** as a **scale factor**. Thus we can restrict our attention to curves given by $r = 1 + k \cos(t)$, or $r = 1 + k \sin(t)$, for various values of the **parameter** k.

EXAMPLE 3. Let r = 1+k cos(t).

First, define r by **Enter**ing the following commands.

Clear[r]

r[t_,k_]:=1 + k Cos[t]

❑ **3.1.** Without a plot, you should be able to describe the graph of r[t,0]. What is it?

❑ **3.2.** Plot r[t,.25], $0 \leq t \leq 2\pi$ and save its graph as **graph1** by **Enter**ing
graph1=PolarPlot[r[t,.25],{t,0,2Pi},AspectRatio->Automatic]. Describe the shape
of the graph.

❑ **3.3.** Plot r[t,.75], $0 \leq t \leq 2\pi$ and save its graph as **graph2**. Describe the shape of the graph.
(Also state the command you **Enter**ed to obtain this result.)

❑ **3.4.** Plot r[t,1.5], $0 \leq t \leq 2\pi$ and save its graph as **graph3**. Describe the shape of the graph.
This is the shape usually identified with the limaçon. (Also state the command you **Enter**ed to
obtain this result.)

❑ **3.5.** Using the **Show** command, you can show all three curves on the same graph. What is the
exact command you would **Enter**? (Be sure to try it!)

❏ **3.6.** **Enter** the command
Do[PolarPlot[r[t,k],{t,0,2Pi},PlotLabel->{"k   ",k}],{k,0,.21,.3}] and carefully
observe the resulting graphs. Then, animate the graphics. (Recall that to animate graphics:
(1) Use the cursor to select the series of graphic cells that you want to animate; and (2) Move the
cursor to Graph and select Animate Selected Graphics.) Based on these results, **and other
graphing you may want to do**, try to determine answers to the following. (You should
note that your answers are **conjectures**. i.e. You haven't **proven** anything!)

(a) What values of k produce a loop?

(b) What values of k produce a "dent"?

(c) What values of k produce an "egg"?

(d) There is a certain value of k, call it **k1**, which has the following property: If one value
of k is taken slightly larger than k1 and another value of k is taken slightly smaller than k1,
then one of the two graphs has a loop and the other one doesn't. Find k1 and then plot r[t,k1].
(The resulting graph is called a **cardioid**.)

(e) State the commands you executed in arriving at the conclusions above.

❏ **3.7.** In the preceding, k≥0. What happens if k<0? Do we get any new shapes that were not
obtained for k≥0? Explain your answer.

❏ **3.8.** For fixed k and a (a≠0), compare the graphs of r=1+k cos(t) and r=a(1+k cos(t)).

❏ **3.9.** Let r=a+b cos(t). Classify the behavior of this two-parameter family into shapes which
depend on a and b. Note: Refer to 3.6.1, 3.6.2, 3.6.3, 3.8, and the remarks preceding
EXAMPLE 3.

Vector Algebra

NAME(S):   **INSTRUCTOR:**
CLASS: **DATE:**

■ INTRODUCTION

The purpose of this lab is to use *Mathematica* to define standard vector algebra operations and use the results to perform elementary computations.

Let $u = \langle u_1, u_2, u_3 \rangle$ be a vector in 3-space. Then the **norm** of u is $\|u\| = \sqrt{u_1^2 + u_2^2 + u_3^2}$.

If u is non-zero, then the three **direction cosines** of u are $\cos\alpha = \dfrac{u_1}{\|u\|}$, $\cos\beta = \dfrac{u_2}{\|u\|}$, and $\cos\gamma = \dfrac{u_3}{\|u\|}$.

Let $v = \langle v_1, v_2, v_3 \rangle$. Then the **dot product** of u and v is $u \cdot v = u_1 v_1 + u_2 v_2 + u_3 v_3$. If u and v are non-zero, then the cosine of the angle θ between them is given by $\cos\theta = \dfrac{u \cdot v}{\|u\| \, \|v\|}$. The **cross product** of u and v is given by $u \times v = \langle u_2 v_3 - u_3 v_2, u_3 v_1 - u_1 v_3, u_1 v_2 - u_2 v_1 \rangle$.

■ NEW *MATHEMATICA* COMMANDS

In this lab, we will define two non-standard (i.e. user-defined) commands **norm** and **cross**. The standard *Mathematica* command **Det** will be used in the definition of cross. **Det** represents the determinant function. You will find the two user-defined commands will relate naturally to the definitions given above.

■ VECTOR ALGEBRA

EXAMPLE 1. Defining the cross product of two vectors in 3-space.

❏ 1.1. Suppose **a** and **b** are vectors in 3-space. We will use *Mathematica* to define their cross product **cross[a,b]**. Carefully type and **Enter** the following commands. Indicate the output and interpret the result of each .
i={1,0,0}

j={0,1,0}

k={0,0,1}

```
cross[a_,b_]:=Det[{{i,j,k},a,b}]
```

```
cross[{a1,a2,a3},{b1,b2,b3}]
```

❏ **1.2.** What is actually produced by the user-defined function **cross[a,b]**?

❏ **1.3.** Carefully type and **Enter** each of the following commands. Indicate the output and interpret the result of each.
```
a={a1,a2,a3}
```

```
b={b1,b2,b3}
```

```
cross[a,b]
```

```
a.b
```

```
a.cross[a,b]
```

```
Simplify[a.cross[a,b]]
```

```
b.cross[a,b]
```

```
Simplify[b.cross[a,b]]
```

EXAMPLE 2. Defining the norm of a vector in 3-space.

❏ **2.1.** Suppose **v** is a vector in 3-space. We will use *Mathematica* to define its norm **norm**
Carefully type and **Enter** the following commands. Indicate the output and interpret
the result of each .
```
norm[v_]:=Sqrt[v.v]
```

```
norm[{v1,v2,v3}]
```

```
norm[v]
```

❏ **2.2.** What is actually produced by the user-defined function **norm[v]**?

EXAMPLE 3. Let $v = -2i + x^2j + 3k = \langle -2, x^2, 3 \rangle$ and $w = -3yi + 4j + 5xk = \langle -3y, 4, 5x \rangle$.

First define **v** and **w** by **Enter**ing the following commands.

v={-2,x^2,3}
w={-3y,4,5x}

❑ **3.1.** If **v** and **w** are perpendicular what relationship(s) exist between them?

❑ **3.2.** Find conditions on x and y so that **v** and **w** are perpendicular. Justify your answer.

❑ **3.3.** If **v** and **w** are parallel what relationship(s) exist between them?

❑ **3.4.** Find conditions on x and y so that **v** and **w** are parallel. Justify your answer.

EXAMPLE 4. Let **u**=2i-j+3k and **v**=i+3j-5k, where **i**, **j**, and **k** are defined in Example 1.

Carefully define **u** and **v** and then calculate each of the following.

❑ **4.1.** $\|v\|$

❑ **4.2.** Find the direction cosines of **u**.

❑ **4.3.** Find $\cos\theta$ where θ is the angle between **u** and **v**.

❏ **4.4.** $u \times v$

❏ **4.5.** $u \cdot (u \times v)$

❏ **4.6.** Find a vector in the same direction as u which has norm 1. Indicate how you set up the problem.

EXAMPLE 5. Let $u = \langle u1, u2, u3 \rangle$ and $v = \langle v1, v2, v3 \rangle$.

Carefully define u and v, then calculate each of the following and simplify your results.

❏ **5.1.** $u \cdot (u \times v)$

❏ **5.2.** $v \cdot (u \times v)$

❏ **5.3.** What conclusion(s) can one reasonably conclude based on the above results?

EXAMPLE 6. The four hydrogen atoms of the methane molecule CH_4 are located at the four vertices of a regular tetrahedron. The distance from the center of each hydrogen atom to the center of the carbon atom is 1.10 angstroms and the angle formed by the line segments connecting a hydrogen atom to the carbon atom and the carbon atom to another hydrogen atom is 109.5 degrees.

❏ **6.1.** Find the distance between two hydrogen atoms.

Parametric Curves

NAME(S):   INSTRUCTOR:
CLASS: DATE:

▪ INTRODUCTION

There is a vast richness of shapes that can be obtained by graphing parametric curves defined by relatively simple functions. The purpose of this lab is to illustrate that richness.

Typically, if x(t) and y(t) are solely defined in terms of sin(t) and cos(t), then the shape obtained from plotting these equations will be a closed figure. In the examples below you will be asked to vary the parameters of a given class of parametric curves and thus try to understand how the shapes of these curves change and, on the other hand, how they remain similar.

▪ NEW *MATHEMATICA* COMMANDS

There are no new *Mathematica* commands used in this lab. However, we will refresh your memory on **ParametricPlot,** a command you have already used. The general form of this command is
                    ParametricPlot[{x[t],y[t]},{t,tstart,tfinish}]
This command is a standard *Mathematica* command and plots the points (x[t],y[t]) for t in the interval **[tstart,tfinish]**. Before the command is **Entered**, x[t], y[t], tstart, and tfinish have to be defined. To see an example from Lab # 25, **Enter** the following commands.
   x[t_]:=Sin[t]
   y[t_]:=Cos[2 t]
   ParametricPlot[{x[t],y[t]},{t,0,2Pi}]
Or, alternatively, you could **Enter** the following:
   f[t_]:={Sin[t],Cos[2 t]}
   ParametricPlot[f[t],{t,0,2Pi}]

NOTE. If you are concerned about the possible distortion of your graph when using **ParametricPlot**, you may want to use the option **AspectRatio->Automatic**

▪ PARAMETRIC CURVES

EXAMPLE 1.   x(t)=k cos(t) + cos(kt)
y(t)=k sin(t) - sin(kt)

For simplicity, let's define **leaf[k,t],** rather than leaf[t], to be the pair (x(t),y(t)) above. (This will make the graphing simpler to handle.) To define **leaf[k,t], Enter** the following commands.

Clear[leaf,x,t]
leaf[k_,t_]:= {k Cos[t] + Cos[k t] , k Sin[t] - Sin[k t]}

❑ **1.1.** Plot **leaf[2,t]** on the interval [0,2π].

❑ **1.2.** Plot **leaf[3,t]** on the interval [0,2π].

❑ **1.3.** Plot **leaf[5,t]** on the interval [0,2π].

❑ **1.4.** Plot **leaf[11,t]** on the interval [0,2π].

❑ **1.5.** Without plotting it, how many cusp points will the graph of **leaf[20,t]** have when plotted on the interval [0,2π]?

❑ **1.6.** Based on what has been done so far, conjecture how many cusp points the graph of **leaf[k,t]** will have when plotted on the interval [0,2π].

❑ **1.7.** The cusp points should appear at the points at which the curve does not possess a derivative. If we plot D[x[k,t],t] for each k, we see that D[x[k,t],t] has more zeros than the curve has cusp points. Why does this seem to be true and how could you explain this behavior? Justify your answer. HINT. Plot **D[x[k,t],t]** and **D[y[k,t],t]** simultaneously for different choices of k.

▣ VARIATIONS ON EXAMPLE 1

EXAMPLE 2.   x(t)=k cos(t) + cos((k-1)t)
y(t)=k sin(t) - sin(kt)

❏ **2.1.** When k=5, graph the curve on the interval [0,2π]. (State all commands you use). Discuss how the shape, etc. is different from the corresponding curve in Example 1.

EXAMPLE 3.   x(t)=k cos(t) + cos(kt)
y(t)=k sin(t) - sin((k-2)t)

❏ **3.1.** When k=5, graph the curve on the interval [0,2π]. (State all commands you use). Discuss how the shape, etc. is different from the corresponding curve in Example 1.

EXAMPLE 4.   x(t)=k cos(t) + cos(kt)
y(t)=k cos(t) - sin(kt)

❏ **4.1.** When k=5, graph the curve on the interval [0,2π]. (State all commands you use). Discuss how the shape, etc. is different from the corresponding curve in Example 1.

Surfaces In 3-Space

NAME(S):   INSTRUCTOR:
CLASS: DATE:

■ INTRODUCTION

One of the many things VERY difficult for us (i.e. humans) to do is to graph anything other than fairly standard functions of a single variable in 2-space or perhaps draw very simple structures in 3-space. We can at best "handwave" at graphing most functions of two variables in 3-space!! Fortunately for us, *Mathematica* has very powerful graphics capabilities. We will certainly use this power to advantage in this lab.

■ NEW *MATHEMATICA* COMMANDS AND OPTIONS

Mathematica provides several commands, with a number of options, that can be particularly useful in understanding the behavior of a function z=f(x,y). **Plot3D** is the only new *Mathematica* command we will use in this lab. However, there are several new options that will be used. An overview of the commands and options to be used in this lab follows. **WARNING. Graphing in 3-space uses a lot of memory. Before executing such a command, save your file!**

■ the graph of z = f(x,y)

The general format for plotting a function z = f(x,y) over a rectangular region is
> **Plot3D[f[x,y],{x,a,b},{y,c,d}]**

This command can be used to plot z = f(x,y) over the rectangular region a ⩽ x ⩽ b, c ⩽ y ⩽ d. For an example, **Enter** the following commands:
Clear[f]
f[x_,y_]:=x^3 -3x y^2
gf1=Plot3D[f[x,y],{x,-3,3},{y,-3,3}]
This graphs f(x,y) over the rectangular region -3 ⩽ x ⩽ 3, -3 ⩽ y ⩽ 3. Notice we have labeled, or named, the graph. Thus, when we want to refer to this graph again, we can do so by referring to its name.

There are several **options** that can be used to obtain a different view. Each option is referred via a name. The general format for a named option is: **OptionName -> OptionValue**
Some standard options for graphs created by **Plot3D** are:

1. **BoxRatios->{scalex,scaley,scalez}**
   This option is used to change the scaling in the (x,y,z) direction. The default is (1,1,.4) . Notice this means the z-direction is scaled to four-tenths its actual value.

The command that can be used to redraw a graph is the **Show** command. Suppose we want to **redraw** the graph of z[x,y] and **rescale** the z-direction to be 1. We can do this by executing the command **Show[gf1,BoxRatios->{1,1,1}]**   (The result is shown below.)

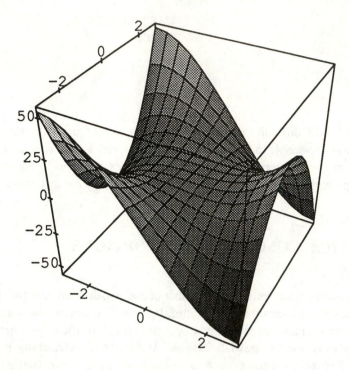

2. PlotPoints->N

This option can be used to plot f[x,y] by evaluating f at more points {x,y} in the xy-plane. The default is 15 points in the x-direction and 15 points in the y-direction. There are two variations on this command. The option **PlotPoints->N** specifies that N points should be used in both the x and y direction. **PlotPoints->{Nx,Ny}** specifies different numbers of sample points for the x and y direction.

Suppose we want to sample the function z[x,y] at 20 points in both the x and y direction. Remember that this will take longer to plot since we are evaluating the function at 20x20 = 400 points instead of the standard of 15x15 = 225 points. We can accomplish this by **Entering** the command **Plot3D[f[x,y],{x,-3,3},{y,-3,3},PlotPoints->20]**

3. ViewPoint->{x,y,z}

This option can be used to view the graph from a different angle. The coordinates {x,y,z} gives the position of the viewpoint relative to the center of the three-dimensional box that contains the object being plotted. The viewpoint is given in a scaled coordinate system in which the longest side of the bounding box has length 1. The center of the bounding box is taken to be {0,0,0}. Some common settings are:

{1.3,-2.4,2}.............default setting,
{0,-2,0}...................directly in front,
{-2,-2,0}..................left-hand corner.

150

Suppose we view the graph of z[x,y] by looking at it from the left-hand corner. We can accomplish this by using the option **Show[gf1,ViewPoint->{-2,-2,0}]**   The resulting graph is shown below.

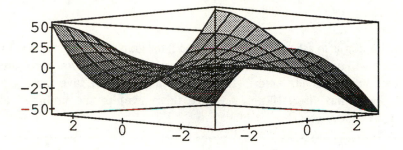

4. Boxed->False

This option can be used to remove the bounding box. For example, **Show[gf1,Boxed->False]** shows gf1 without the bounding box.

■ the contour lines of z= f(x,y)

The general format for plotting the contours of a function z = f(x,y) over the rectangular region $a \leqslant x \leqslant b$, $c \leqslant y \leqslant d$ is

$$ContourPlot[f[x,y],\{x,a,b\},\{y,c,d\}]$$

ContourPlot returns a graph of f(x,y) = zconstant for different values of zconstant. In essence this depicts the curves obtained by slicing the surface z = f(x,y) with a plane parallel to the xy-plane and at a distance zconstant from the xy-plane. The two most useful options for this command are:

1. Contours->N (In version 1, this option is ContourLevels->N)

This option chooses N equally spaced contours between the minimum and maximum z values.

2. PlotPoints->N

This option specifies the number of points in each direction at which to sample the function.

Enter the command **ContourPlot[f[x,y],{x,-3,3},{y,-3,3}]** to obtain a plot of the contours of f[x,y]. Is this what you expected to see?

151

■ **the cross-sectional view obtained by slicing the surface z = f(x,y) with a plane perpendicular to the xy-plane or the yz-plane.**

If we wish to look at the cross-sectional view obtained by slicing the surface z = f(x,y) with the plane x=2, a plane parallel to the yz-plane which is 2 units from the origin, **Enter** the command **Plot[f[2,y],{y,-3,3}]**

To slice the surface with a plane y=2, a plane parallel to the xz-plane and 2 units from the origin, **Enter** the command **Plot[f[x,2],{x,-3,3}]**

■ **EXAMPLES**

It would not be practical for us to ask you to sketch (by hand) the graphs requested on the following examples. Thus, space will not be provided for this. However, we do expect you to construct the graphs and turn them in to your instructor. **If you have a a printer, save these graphs on another file and print that file. If not, be sure these graphs are included on the disk you turn in to your instructor.**

EXAMPLE 1. $f(x,y) = \sin\left(\dfrac{xy}{1+x^2+y^2}\right)$

❑ **1.1.** Plot f over an appropriate region. Use **PlotPoints->30** to give a better picture.

❑ **1.2.** Obtain a contour plot of f over the same region.

❑ **1.3.** Sketch the curves obtained by slicing the curve with the planes x=-2, x=-1, x=0, x=1, and x=2.

❑ **1.4.** Sketch the curves obtained by slicing the curves with the planes y=-2, y=-1, y=0, y=1, and y=2.

❑ **1.5.** Observe what happens when x=0 or y=0.

EXAMPLE 2. $f(x,y) = \dfrac{\sin(xy)}{xy}$

❑ **2.1.** Plot f over an appropriate region.

❑ **2.2.** Obtain a contour plot of f over the same region.

❑ **2.3.** Sketch the curves obtained by slicing the curve with the planes x=-2, x=-1, x=0, x=1, and x=2.

❑ **2.4.** Sketch the curves obtained by slicing the curves with the planes y=-2, y=-1, y=0, y=1, and y=2.

❑ **2.5.** Observe what happens when x=0 or y=0.

EXAMPLE 3. $f(x,y) = e^{-x^2} + e^{-4y^2}$

❑ **3.1.** Plot f over an appropriate region. Use **BoxRatios->{1,1,1}** to give a better perspective. Also, redraw the graph using the **Show** command and changing the viewpoint to **ViewPoint->{2,.5,2}**

❑ **3.2.** Obtain a contour plot of f over the same region.

❑ **3.3.** Sketch the curves obtained by slicing the curve with the planes x=-2, x=-1, x=0, x=1, and x=2.

❑ **3.4.** Sketch the curves obtained by slicing the curves with the planes y=-2, y=-1, y=0, y=1, and y=2.

❑ **3.5.** Observe what happens when x=0 or y=0.

EXAMPLE 4. $f(x,y) = \dfrac{xy^3 - x^3y}{2}$

❏ **4.1.** Plot f over an appropriate region. Redraw the graph using the **Show** command and changing the viewpoint to **ViewPoint->{2,0,1}**

❏ **4.2.** Obtain a contour plot of f over the same region.

❏ **4.3.** Sketch the curves obtained by slicing the curve with the planes x=-2, x=-1, x=0, x=1, and x=2.

❏ **4.4.** Sketch the curves obtained by slicing the curves with the planes y=-2, y=-1, y=0, y=1, and y=2.

❏ **4.5.** Observe what happens when x=0 or y=0.

154

Partial Derivatives

NAME(S):   INSTRUCTOR:

CLASS: DATE:

▣ INTRODUCTION

The graph of $y = f(x)$, in general, is a curve in the xy-plane. The graph of $z = f(x,y)$, in general, is a surface in xyz-space. When f is a function of only one variable, the derivative of the function $y = f(x)$ is given by the limit

$$\lim_{h \to 0} \frac{f(x+h) - f(x)}{h}$$

which measures the rate of change of $f(x)$ with respect to the change in x. For functions of two or more independent variables, the calculation of a rate of change of $f(x,y)$ at $(x0,y0)$ using limits becomes more involved since "nearby" points of $(x0,y0)$ can approach $(x0,y0)$ along infinitely many paths. In order to get a "handle" on defining a derivative of a function $f(x,y)$ of two independent variables, we start by examining rates of change along paths that are parallel to each of the coordinate axes. This approach leads to determining the partial derivative of f with respect to x if the path is one that is parallel to the x-axis, and the partial derivative with respect to y if the path is parallel to the y-axis. There are two approaches that one can take in defining these partial derivatives, one algebraic and the other geometrical.

■ The algebraic approach

If we restrict the point (x,y) to lie on the line through $(x0,y0)$ that is parallel to the x-axis then we are restricted to considering points of the form $(x,y0)$. In this case the partial derivative of f with respect to x is given by the limit

$$\frac{\partial f}{\partial x}(x0,y0) = \lim_{h \to 0} \frac{f(x0+h, y0) - f(x0,y0)}{h}.$$

Similarily, the partial derivative of f with respect to y is given by the limit

$$\frac{\partial f}{\partial y}(x0,y0) = \lim_{k \to 0} \frac{f(x0,y0+k) - f(x0,y0)}{k}.$$

We note that determining the partial derivatives with respect to x (or y) can be viewed as differentiating the function $z = f(x,y)$ with respect to x (or y) while holding the y (or x) variable fixed. Thus the rules for differentiating functions of a single variable apply. In *Mathematica* the partial derivatives can be determined by using the **D** operator as follows.

155

Clear[f]
f[x_,y_]:= x^2 + Sin[x y]   (* **Define a function of two variables** *)

To find $\dfrac{\partial f}{\partial x}$, which we will label as **fx** , one **Enters fx = D[f[x,y],x]**

To find $\dfrac{\partial f}{\partial y}$, which we will label as **fy,** one **Enters fy = D[f[x,y],y]**

In a similar manner, the partial derivative of each of these functions can be obtained using the **D**

operator. Thus, **fxx = D[fx,x]** will give $\dfrac{\partial^2 f}{\partial x^2}$, the partial derivative of fx with respect to x. Likewise,

fyy=D[fy,y]   (* **the partial derivative of fy with respect to y***)
fxy=D[fx,y]   (* **the partial derivative of fx with respect to y** *)

The second partials, $\dfrac{\partial^2 f}{\partial x^2}$, $\dfrac{\partial^2 f}{\partial y^2}$, and $\dfrac{\partial^2 f}{\partial x \partial y}$, can also be obtained by **Entering**

fxx=D[f[x,y],{x,2}]
fyy=D[f[x,y],{y,2}]
fxy=D[f[x,y],x,y]

■ **The geometric approach**

To interpret fx geometrically, we note that the set of points (x,y0,z) is the plane in xyz-space that is
perpendicular to the xy-plane and passes through (x0,y0,f(x0,y0)). The intersection of this plane with
the graph of f(x,y) will be a curve in xyz-space. We will label this curve as **xcurve(t)**. The curve
xcurve(t) is called the **lift** of the line xline(t) onto the surface z=f(x,y). Geometrically, the curve
xcurve(t) represents the path of z=f(x,y) when x and y are restricted to lie on the line through (x0,y0)
that is parallel to the x-axis. Thus, the slope of the tangent line to this curve at the point
(x0,y0,f(x0,y0)), if it exists, will represent fx. Similarly, the intersection of the plane x=x0 with the
graph of f(x,y) is a curve in xyz-space and the slope of this curve, if it exists, is fy. We label this curve
as **ycurve(t)**. Symbolically,

xcurve(t) = Plane(x = x0) ⋂ graph(f(x,y)) and **ycurve(t)** = Plane(y = y0) ⋂ graph(f(x,y)).

We obtain the equation for xcurve(t) by first finding the equation of the line lying in the x-axis and then
obtain xcurve(t) by restricting f to be defined only at points along this line. Thus, we need to:

1. Determine the equation of the line xline(t) that lies in the xy-plane, is parallel to the x-axis, and that
   passes through the point P=(x0,y0). Thus, we proceed as follows.

   Clear[f]
   f[{x_,y_}]:=-(x^2+y^2) +1   (* **Define f(x,y)** *)
   x0=1
   y0=-1

```
P={x0,y0}                    (* the point (x0,y0) *)
xvector={1,0}                (* a vector parallel to the x-axis *)
xline[t_]:=P + t xvector
```

2. Apply f to this set of points.

```
Clear[xcurve]
xcurve[t_]:=Flatten[{xline[t],f[xline[t]]}]
     (* xcurve[t] is the lift of xline[t] onto the surface z=f(x,y) *)
```

We have to use the **Flatten** command to eliminate all but the outermost parentheses.

■ NEW *MATHEMATICA* COMMANDS

Two non-standard commands have been created to graph the curves xcurve(t) and ycurve(t). They are:
1. **tanplot[f, theta, {x0,y0}, {x,a,b}, {y,c,d}]**, and
2. **curvetnplot[f,{x0,y0},thetaset]**.
These commands are available from your instructor.

The command **tanplot[f, theta, {x0,y0}, {x,a,b}, {y,c,d}]** will plot the graph z=f(x,y) and the curve curve(t) obtained by restricting f to lie on the line through the point {x0,y0} that has been rotated theta degrees with respect to the positive x-axis. If f is restricted to lie along a line parallel to the x-axis then theta = 0, and if f is restricted to lie along a line parallel to the y-axis then theta will be $\pi/2$.
The **parameters** in the command tanplot are:
 (a) **f** ....the function z = f(x,y),
 (b) **theta**........................the angle of inclination of the line used to determine the curve curve(t),
 (c) **{x0,y0}**......................the x-y coordinates of the point of tangency,
 (d) **{x,a,b}** and **{y,c,d}**....the region over which the surface z= f(x,y) will be plotted.
 CAUTION. Since this command plots a 3-dimensional surface, executing this command uses a lot of memory! Make sure you save your file before execution!!!

The command **curvetnplot[f,{x0,y0},thetaset]** will plot the curves and their tangent lines determined by restricting f to lie along the lines in the xy-plane whose angles of inclination are specified in the set thetaset.
The **parameters** in the command curveplot are:
 (a) **f**....................the function z = f(x,y),
 (b) **{x0,y0}**..........the x-y coordinates of the point of tangency,
 (c) **thetaset**..........a set {theta1 , theta2 , theta3 , etc} of angles. Each angle represents the inclination of a line lying in the xy-plane. For each line, the curve obtained by restricting the points (x,y) to lie on this line is plotted in xyz-space.

Both these commands have an option, **time->{t0,t1}**, that can be used to specify the range of t values over which the curve is plotted. The default value for t is -1 ⩽ t ⩽ 1.

NOTE. To use these commands, f must be defined as a vector function. (i.e. define f using f[{x_,y_}]:= - - -.). Enclosing the pair (x,y) in braces rather than parentheses makes f a function defined on a vector, the vector {x,y}.

■ PARTIAL DERIVATIVES

> **EXAMPLE 1.** $f(x,y) = x^2 + y^2$ and $P0 = \{-1,1\}$.

First, define f and P0 by **Entering** the following commands.
Clear[f,xcurve,ycurve]
f[{x_,y_}]:=x^2 + y^2 + 1
P0={1,-1}

❏ **1.1.** What is the equation of the curve obtained by restricting f to lie on the line in the xy-plane that is parallel to the x-axis and passes through the point P0?

❏ **1.2.** What is the equation of the curve obtained by restricting f to lie on the line in the xy-plane that is parallel to the y-axis and passes through the point P0?

❏ **1.3.** Sketch each of the curves from questions 1.1 and 1.2. by **Entering**
curvetnplot[f,P0,{0,Pi/2},time->{-2,2}]

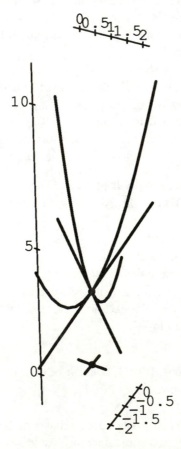

❑ **1.4.** What is the slope of the tangent line to the curve xcurve(t) at the point P0?

❑ **1.5.** What is the slope of the tangent line to the curve ycurve(t) at the point P0?

❑ **1.6.** Sketch the surface z = f(x,y) and show the curve xcurve(t) admits a tangent line at the point P0 by **Enter**ing the following commands.
theta=Pi/2
tanplot[f,theta,P0,{x,-2,2},{y,-2,2},time->{-1.5,1.5}]
NOTE. This graph will take a while to plot, maybe a couple of minutes. Also, you might want to enlarge the picture by selecting the graph and then expanding it by dragging the lower right-hand corner.

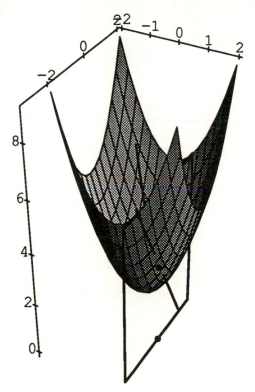

EXAMPLE 2. $f(x,y) = x^2 - y^2$ and $P0 = \{-1,1\}$.

❑ **2.1.** What is the equation of the curve obtained by restricting f to lie on the line in the xy-plane that is parallel to the x-axis and passes through the point P0?

❑ **2.2.** What is the equation of the curve obtained by restricting f to lie on the line in the xy-plane that is parallel to the y-axis and passes through the point P0?

❑ **2.3.** Sketch each of the curves from question 2.1 and 2.2.

❑ **2.4.** What is the slope of the tangent line to the curve xcurve(t) at the point P0?

❑ **2.5.** What is the slope of the tangent line to the curve ycurve(t) at the point P0?

❏ **2.6.** Sketch the surface z = f(x,y) and show the curve xcurve(t) and its tangent line at the point P0.

EXAMPLE 3. $f(x,y) = \sin(x) \cos(y)$ and $P0 = \{-1, 1\}$.

❏ **3.1.** What is the equation of the curve obtained by restricting f to lie on the line in the xy-plane that is parallel to the x-axis and passes through the point P0?

❏ **3.2.** What is the equation of the curve obtained by restricting f to lie on the line in the xy-plane that is parallel to the y-axis and passes through the point P0?

❏ **3.3.** Sketch each of the curves from question 3.1 and 3.2.

161

❏ **3.4.** What is the slope of the tangent line to the curve xcurve(t) at the point P0?

❏ **3.5.** What is the slope of the tangent line to the curve ycurve(t) at the point P0?

❏ **3.6.** Sketch the surface z = f(x,y) and show the curve xcurve(t) and its tangent line at the point P0. HINT. Plot the surface over the region determined by $-\pi/2 < x < \pi/2$ and $-\pi/2 < y < \pi/2$.

Directional Derivatives

NAME(S):   **INSTRUCTOR:**
CLASS: **DATE:**

▪ INTRODUCTION

As mentioned previously, the graph of z = f(x,y), in general, is a surface in xyz-space. As we saw in LAB # 30, the partial derivatives can be interpreted as determining the rates of change of f(x,y) as x(or y) varies along a line that is parallel to one of the axes, either the x-axis or the y-axis. An obvious generalization of this approach is to determine the rates of change of z = f(x,y) as the point (x,y) in the xy-plane varies along a straight line, but not necessarily a line that is parallel to one of the axes. This approach leads to determining a directional derivative. As with partial derivatives there are two approaches one can take in defining directional derivatives, one algebraic and the other geometrical.

■ The algebraic approach

Suppose we restrict the points (x,y) to lie on a slanted line through the point **xbar**=(x0,y0). If we assume that the angle of inclination of this line(with respect to the positive x-axis) is theta, then the line must satisfy the equation x(t) = **xbar** + t**v** where **v** = (cos(theta),sin(theta)). Again, considering the rate of change of f(x,y) along a line as the limiting position of the secant line formed by two nearby points, it seems natural to define the directional derivative of f(x,y) at **xbar** in the direction **v** to be

$$D_v f(\textbf{xbar}) = \lim_{t \to 0} \frac{f(\textbf{xbar} + t\,v) - f(\textbf{xbar})}{t} \quad \text{where} \quad \textbf{xbar} = (x0, y0) \text{ and } v = (\cos(\text{theta}), \sin(\text{theta})).$$

■ The geometrical approach

Let's see if we can interpret geometrically the terms that appear in the definition of the directional derivative

$$D_v f(\textbf{xbar}) = \lim_{t \to 0} \frac{f(\textbf{xbar} + t\,v) - f(\textbf{xbar})}{t} \quad \text{where} \quad \textbf{xbar} = (x0, y0) \text{ and } v = (\cos(\text{theta}), \sin(\text{theta})).$$

Geometrically, the equation **xbar** + t**v** represents a line in the xy-plane that passes through the point **xbar** and has direction **v**. Therefore, f(**xbar** + t**v**) represents the lift of this line onto the surface z=f(x,y). This can equivalently be interpreted as the curve obtained when the surface z=f(x,y) is intersected with the plane that is perpendicular to the xy-plane and intersects the xy-plane in the line **xbar** + t**v**. We shall label this curve as vcurve(t). With this interpretation, the directional derivative becomes the slope of the tangent line to the curve vcurve(t) at the point **xbar**.

▣ NEW *MATHEMATICA* COMMANDS

Two non-standard commands from LAB # 30 are also available for your use in this lab. They are:
1. **tanplot[f, theta, {x0,y0}, {x,a,b}, {y,c,d}]**, and
2. **curvetnplot[f,{x0,y0},thetaset]**.
These commands are available from your instructor.

The command **tanplot[f, theta, {x0,y0}, {x,a,b}, {y,c,d}]** will plot the graph z=f(x,y) and the curve curve(t) obtained by restricting f to lie on the line through the point {x0,y0} that has been rotated theta degrees with respect to the positive x-axis. If f is restricted to lie along a line parallel to the x-axis then theta = 0, and if f is restricted to lie along a line parallel to the y-axis then theta will be Pi/2. The **parameters** in the command tanplot are:

(a) **f** ....the function z = f(x,y),

(b) **theta**..........................the angle of inclination of the line used to determine the curve curve(t),

(c) **{x0,y0}**......................the x-y coordinates of the point of tangency,

(d) **{x,a,b}** and **{y,c,d}**....the region over which the surface z= f(x,y) will be plotted.

CAUTION. Since this command plots a 3-dimensional surface, executing this command uses a lot of memory! Make sure you save your file before execution!!!

The command **curvetnplot[f,{x0,y0},thetaset]** will plot the curves and their tangent lines determined by restricting f to lie along the lines in the xy-plane whose angles of inclination are specified in the set thetaset.
The **parameters** in the command curveplot are:

(a) **f**.....................the equation of the function z = f(x,y),

(b) **{x0,y0}**..........the x-y coordinates of the point of tangency,

(c) **thetaset**..........a set {theta1 , theta2 , theta3 , etc} of angles. Each angle represents the inclination of a line lying in the xy-plane. For each line, the curve obtained by restricting the points (x,y) to lie on this line is plotted in xyz-space.

Both these commands have an option, **time->{t0,t1}**, that can be used to specify the range of t values over which the curve is plotted. The default value for t is -1 ≤ t ≤ 1.

NOTE. To use these commands, f must be defined as a vector function. (i.e. define f using f[{x_,y_}]:= - - -.). Enclosing the pair (x,y) in brackets rather than parentheses makes f a function defined on a vector, the vector {x,y}.

To see an example of the graph obtained by using curvetnplot, **Enter** the following commands.
Clear[f]
f[{x_,y_}]:=-x^2+y^2+1
curvetnplot[f,{0,0},{Pi/2,Pi/3}]

164

An illustrated example of the output from the command **tanplot** is shown below.

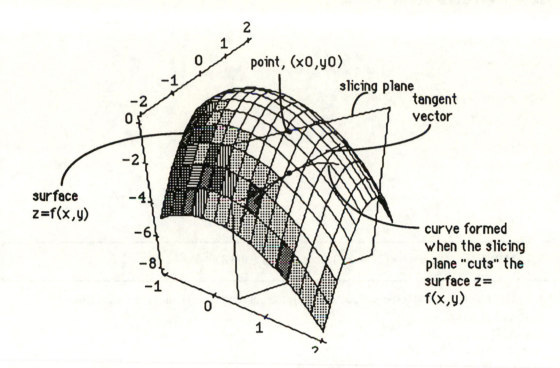

▣ DIRECTIONAL DERIVATIVES

EXAMPLE 1. $f(x,y) = x^2 + 4y^3$, xbar $= (-1,1)$, and $\theta = \dfrac{\pi}{3}$.

❏ **1.1.** Find the directional derivative of f at xbar in the direction v=(cos(θ),sin(θ)) by using the definition of the directional derivative. (i.e. use limits.)

HINT. Let's first define f, xbar, θ, and the direction vector v by **Enter**ing the following commands.

Clear[f]
f[{x_,y_}]:=x^2 + 4 y^3
theta=Pi/3
xbar={-1,1}
v={Cos[theta],Sin[theta]}
Now, find the directional derivative of f at xbar in the direction v by **Enter**ing
dv=Limit[(f[xbar + t v] - f[xbar])/t,t->0]

165

❏ **1.2.** Find the unit vector u for which the directional derivative $D_u f(\{-1,1\})$ is maximum. State the steps you take as well as the answer.

HINT. Find an expression for D_u in terms of theta. Then use derivative information to find the maximum point.

EXAMPLE 2. $f(x, y) = -(x^2 + y^2)$, xbar = $(1, -1)$, and $\theta = \dfrac{\pi}{3}$.

❏ **2.1.** Use tanplot to graph the surface z=f(x,y), the curve vcurve(t), and the tangent line to vcurve(t) at **xbar**.

Clear[f]
f[{x_,y_}]:=-(x^2+y^2)
tanplot[f,Pi/3,{1,-1},{x,-2,2},{y,-2,2}]

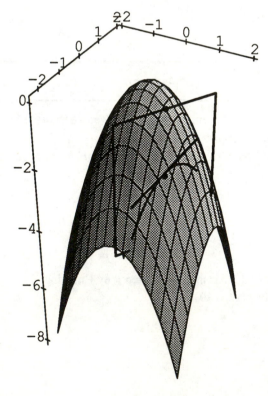

❑ **2.2.** Use curvetnplot to plot the curve vcurve[t] and its tangent line at the point **xbar** for the directions θ=π/4 and θ=π/3. HINT. Consider **curvetnplot[f,{1,-1},{Pi/4,Pi/3}]**

❑ **2.3.** Use curvetnplot to plot the curve vcurve[t] and its tangent line at the point **xbar** for the directions θ=0, θ=π/4, θ=π/3, and θ=π/2. Notice that all the tangent lines seem to lie in a single plane, namely, the tangent plane. Why is this true?

EXAMPLE 3. $f(x,y) = \dfrac{x^4 + y^4 - 4xy}{10}$

Carefully define f as a function of a vector. (i.e. f[{x_,y_}]:=... .)

❑ **3.1.** Find the directional derivative, Dv, of f(x,y) using the limit process for **xbar** = (1.5,1) and θ = -π/6.

❑ **3.2.** Find the gradient of f, denoted grad(f), at **xbar**. Recall that

$$grad(f) = \left(\frac{\partial f}{\partial x}(x0,y0), \frac{\partial f}{\partial y}(x0,y0) \right).$$

❑ **3.3.** Find grad(f)·v (the dot product), where v is the vector determined by 3.1.

❑ **3.4.** What is the relationship between the answer to 3.2. and 3.3.? Under what circumstances will this relationship hold in general?

❑ **3.5.** Use tanplot to graph the surface z=f(x,y), the curve vcurve(t), and the tangent line at **xbar**.

❑ **3.6.** Use curvetnplot to plot the curve vcurve(t) and its tangent line at the point **xbar** for the directions θ=-π/6, θ=0, and θ=π/2.

❏ **3.7.** What is the equation of the curve vcurve(t) at **xbar** when θ = - π/4? When θ = - π/6?

❏ **3.8.** What is the equation of the tangent line to the curve vcurve(t) at **xbar** when θ = - π/6?

EXAMPLE 4.   f(x,y) = sin(x) sin(y)

❏ **4.1.** Find the directional derivative, Dv, of f at **xbar**=(π/2,π/2) for θ1 =- π/4 and θ2=π/4. (Remember that you have two ways at your disposal.)

❏ **4.2.** What is the equation of the curve vcurve(t) when θ = - π/4? When θ= π/4?

❏ **4.3.** What is the equation of the tangent line at P to the curve vcurve(t) when θ = - π/4? When θ = π/4?

❏ **4.4.** Execute the following commands individually and then show the graph obtained from executing the **Show** command. (NOTE. This is VERY memory hungry, so the graph is indicated below.)

gf1=tanplot[f,-Pi/4,{Pi/2,Pi/2},{x,0,Pi},{y,0,Pi}]
gf2=tanplot[f,Pi/4,{Pi/2,Pi/2},{x,0,Pi},{y,0,Pi}]
Show[{gf1,gf2}]

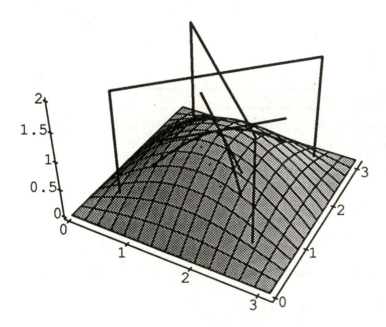

❏ **4.5.** Use curvetnplot to plot the curve vcurve[t] and its tangent line at the point **xbar** for the directions θ =- π/4, θ = 0, and θ = π/2.

Critical Points

NAME(S):   INSTRUCTOR:
CLASS: DATE:

▣ INTRODUCTION

When trying to find the maximum (or minimum) value of a function of one real variable (f: R → R), we were able to cut down the number of points at which the maximum (or minimum) might occur by restricting attention to the critical points. With real-valued functions of two or more variables, there is a similar set of points to which we can restrict our attention if we are trying to determine the local maximum or minimum points of the function. As in the one variable case, this collection of points is called the set of critical points. More formally, a pair (a,b) is a critical point of f if either

   (i) $f_x(a,b) = 0$ and $f_y(a,b) = 0$, or

   (ii) $f_x(a,b)$ or $f_y(a,b)$ does not exist.

For functions of two variables, $z = f(x,y)$, there is a test similar to the second derivative test for one variable. This test involves testing the discriminant $D = f_{xx} f_{yy} - f_{xy}^2$ of f at the critical points for which f and its second partial derivatives are continuous in a "small" open disc about the critical point. This test is:

   Suppose $f_x(a,b) = 0$ and $f_y(a,b) = 0$.

   If $D(a,b) > 0$, then

      (i) if $f_{xx}(a,b) < 0$ then (a,b) is a local maximum point of f,

      (ii) if $f_{xx}(a,b) > 0$ then (a,b) is a local minimum point of f.

   If $D(a,b) < 0$, then (a,b) is a saddle point of f.

In this lab we will examine how we can use the power of *Mathematica* to more efficiently handle finding the critical points and evaluating the function at each critical point in order to determine the local behavior of the function at that critical point.

▣ NEW *MATHEMATICA* COMMANDS

There are no new *Mathematica* commands used in this lab. However, we will mention **ContourPlot** since we have not used it for "awhile". We will be drawing contour plots in a neighborhood of each critical point. A contour plot is essentially a topographic map of the function. The contours join points on the surface that have the same height. The general form you will need to use is
   ContourPlot[f[x,y],{x,a,b},{y,c,d},Contours->n]
The option **Contours->n** specifies the number of contours and has default 10. If you are using Version 1 of *Mathematica*, then you will need to replace **Contours->n** by **ContourLevels->n**

As an example, the command **ContourPlot[x^2+y^2,{x,-5,5},{y,-5,5},Contours->20]** draws a set of contours of f(x,y)=x^2+y^2 about (x,y) = (0,0). We should see a set of concentric circles since (0,0) is a minimum point of the function. We have included the plot option **Contours->20** to enable us to see more contours.

■ PROCEDURE FOR FINDING (AND LABELING) CRITICAL POINTS

Let f(x,y)=.........

Objective: Find all extrema and saddle points.

❏ *1.* *Define the function.*

   Clear[f]
   f[x_,y_]:=.......

❏ *2.* *Find and label all first partial derivatives and second partial derivatives.*

   fx=D[f[x,y],x]
   fy=D[f[x,y],y]
   fxx=D[f[x,y],{x,2}]
   fyy=D[f[x,y],{y,2}]
   fxy=D[f[x,y],x,y]

   OBSERVATION. If f and all of its partial derivatives of all orders are continuous, the second derivative test will apply. In particular, if f is a polynomial in x and y, the second derivative test applies.

❏ *3.* *Find and label the discriminant.*

   disc=fxx fyy - fxy^2

❏ *4.* *Find all critical points.*

   ColumnForm[Solve[{fx==0,fy==0},{x,y}]]

   NOTE. The command **ColumnForm[....]** will return the answer in column form.

❏ *5.* *Evaluate the discriminant at each qualified critical point (a,b) .*

   disc/.x->a/.y->b

❏ *6.* *Evaluate fxx at (a,b) if necessary.*

   fxx/.x->a/.y->b

❏ *7.* *It's decision time!*

■ CRITICAL POINTS

EXAMPLE 1. $f(x,y) = \dfrac{1}{3}x^3 + 4xy - 9x - y^2$

In this case, the second derivative test applies for each critical point and the critical points are isolated. After carefully defining the function, answer the following.

❏ **1.1.** Find all first partials.

❏ **1.2.** Find all second partials.

❏ **1.3.** Find the discriminant.

❏ **1.4.** Find all critical points and determine the type of each. Justify your answer.

❏ **1.5.** Draw contour lines about each of the critical points. Make the x-interval and y-interval 6 units wide and "center" the critical point. Include the option **Contours->20**
NOTE. If you have a printing capability, you can save this plot in a separate file to print after obtaining all the graphs in this session. Otherwise, make sure that each contour plot is included on the disc you turn in to your instructor.

EXAMPLE 2. $f(x,y) = x^{\frac{2}{3}} + y^{\frac{2}{3}} + 1$

In this case, the second derivative test may not apply for each critical point and the critical points may not be isolated. After carefully defining the function, answer the following.

❏ **2.1.** Find the set of critical points. Justify your answer.
HINT. This is **not** a finite set. Do a **Plot3D[......]** to get some idea of the behavior of f.

❑ **2.2.** Construct a contour plot of f about (x,y) = (0,0).
NOTE. If you have a printing capability, you can save this plot in a separate file to print after obtaining all the graphs in this session. Otherwise, make sure that this contour plot is included on the disc you turn in to your instructor.

❑ **2.3.** What is the image of f along lines that are perpendicular to the x-axis? Along lines that are perpendicular to the y-axis?
Some questions to ask that might be helpful.
• What is the set of points {(1,y) | y is any real number}?
• What is the set of points {(x,1) | x is any real number}?

❑ **2.4.** What set of points is plotted by the command **Plot[f[1,y],{y,-4,4}]** ?
NOTE. If you have a printing capability, you can save this plot in a separate file to print after obtaining all the graphs in this session. Otherwise, make sure that this plot is included on the disc you turn in to your instructor.

❑ **2.5.** Describe the set of minimum points, if any, of f.

❑ **2.6.** Describe the set of maximum points, if any, of f.

EXAMPLE 3. $f(x,y) = (x^2 + 3y^2)\, e^{-(x^2 + y^2)}$

❑ **3.1.** Find the set of critical points.
HINT. There are five critical points. Do a **Plot3D[.....]** to get some idea of the behavior of f.

❑ **3.2.** Construct a contour plot about each critical point. For each contour plot, restrict the region in the xy-plane to exclude all but one of the critical points.
NOTE. If you have a printing capability, you can save this plot in a separate file to print after obtaining all the graphs in this session. Otherwise, make sure that these contour plots are included on the disc you turn in to your instructor.

❑ **3.3.** Classify each critical point as either a local maximum point, a local minimum point, or a saddle point.

Applied Max/Min Problems
(several variables)

NAME(S):   INSTRUCTOR:
CLASS: DATE:

▪ INTRODUCTION

In this lab, we will apply the techniques of calculus to find the maximum or minimum values of a function. (You may want to refer to LAB # 32 Critical Points.) Problems of this type fall into two categories.

1. One category deals with finding all local maximum and/or minimum points of a given function. Our approach to problems of this type involves finding all the critical points of the function and then evaluating the function at each of these points.

2. The other category deals with finding the absolute maximum or minimum value of a function in the case where the independent variables are restricted to lie within some known range. If the region of interest is a closed, bounded set and the function is continuous, then we know there is a maximum point and a minimum point. These type problems involve solving two separate subproblems.

One subproblem involves finding the maximum or minimum points that lie within the region of interest. This is similar to the problems of category one except in this case we are only interested in the critical points that lie within the region of interest.

The other subproblem involves finding the maximum or minimum value of the function on the boundary of the region. This typically is a curve that can be expressed by a parametric equation. Problems of this type can be handled by applying the techniques of calculus to functions of a single variable. EXAMPLE 1 is a problem of this type.

▪ NEW *MATHEMATICA* COMMANDS

A command that we can use to help us to understand the behavior of a function f(x,y) is the **density plot**. A density plot will be light over the part of the region where the function has "large" values and will be dark over the part of the region where the function has "small" values.   The general form of the **DensityPlot** command is **DensityPlot[f[x,y], {x,a,b}, {y,c,d}]**
NOTE. The option **PlotPoints->n** can be used. The default is **PlotPoints->15**

■ MAX/MIN PROBLEMS

> **EXAMPLE 1.** A circular metal plate of radius 1 meter is placed with its center at the origin of the xy - plane and heated so that its temperature **temp** at the point (x, y) is given by the formula temp(x, y) = $64(3x^2 - 2xy + 3y^2 + 2y + 5)$, where temp is measured in degrees Celsius.

In this problem, we are interested in determining the maximum temperature, and the minimum temperature, on the plate. *Mathematica* has more than one tool that we can use to help us understand the behavior of a function like the temperature function described above. One tool that we can use is to plot the function over its circular region. One way to effectively do this is by restricting the definition of **temp** when we define it. Define the function **temp** by **Enter**ing
temp[x_,y_]:= 64(3x^2 - 2 x y + 3 y^2 + 2y + 5)/; 0<=x^2 + y^2 <= 1
temp[x_,y_]:= 0 /; x^2 + y^2 >1

To plot **temp** over an appropriate region, we **Enter** the following command and behold the resulting graph!

Plot3D[temp[x,y], {x,-1,1}, {y,-1,1}, PlotPoints->30, Boxed->False]

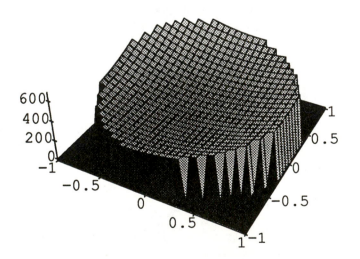

❏ **1.1.** Another tool that we can use to understand the temperature function temp is a contour plot. **Enter** the following and indicate the result.
ContourPlot[temp[x,y], {x,-1,1}, {y,-1,1}, Contours ->30, PlotPoints ->30]
(If you are using Version 1 of *Mathematica*, use **ContourLevels ->30**)

❑ **1.2.** Show the density plot of temp(x,y) by **Entering**
DensityPlot[temp[x,y], {x,-1,1}, {y,-1,1}, PlotPoints ->30]

Look back at the three different "pictures" of the function temp(x,y) and see how each gives you a look at what the function is doing. You should be able to see the "hot spots" and the "cool spots" in the density plot. Can you?

The second derivative test applies for each critical point since temp(x,y) is a polynomial in x and y.

❑ **1.3.** Find all first partials.

❑ **1.4.** Find all second partials.

❑ **1.5.** Find the discriminant.

❑ **1.6.** Find all critical points and determine the type of each. Justify your answer.

❏ **1.7.** What are the critical points of the function temp(x,y) that lie in the interior of the circular metal plate? State the type of each.

❏ **1.8.** What is the maximum and minimum temperature value on the boundary?
HINT. What is the parametric equation for a circle? Execute the following commands and state the result of each.
x=Cos[t]
y=Sin[t]
temp[x,y]

tempt=D[temp[x,y],t]

ans=tempt/.Sin[t]^2->1-Cos[t]^2

Factor[ans]

Minimum temperature on the boundary is _____ .
Maximum temperature on the boundary is _____ .

❏ **1.9.** What is the lowest temperature on the circular plate?

What is the highest temperature on the circular plate?

> **EXAMPLE 2.** A certain state plans to supplement its state budget by selling weekly lottery tickets. Opinion polls show a potential 1 million tickets will be purchased per week at $1 per ticket, but 130,000 fewer tickets will be purchased per week for every $0.25 increase in the price per ticket. Fixed costs such as printing and distributing tickets, salaries of lottery officials, and advertising are expected to run $140,000 per week. Regardless of the price of a ticket, it is estimated that an additional dollar spent in advertisement will result in the sale of one additional ticket per week. The state, by law, must return one-third of its weekly revenue from ticket sales as prizes to the purchasers.

❏ **2.1.** How much should the state charge for a lottery ticket in order to maximize its profits?

❑ **2.2.** What maximum weekly profit from the lottery can be expected?

❑ **2.3.** Suppose that only 120,000 fewer tickets will be purchased for each quarter increase in the price of the ticket. How would this affect the charge that would result in the maximum profit?

❑ **2.4.** Suppose that each additional dollar spent in advertisement will result in the sale of two additional tickets. How would this affect the charge that would result in the maximum profit?

EXAMPLE 3. Approximation of data by a straight line

A typical problem that arises in many situations involves finding the "best" straight line that fits a given set of points $P = \{(x_1, y_1), (x_2, y_2), ..., (x_n, y_n)\}$. One way to define the line $y = mx + b$ is the **Method of Least Squares**. As the following illustration shows, the vertical distance of the data point (x_j, y_j) to the line $y = mx + b$ is

$$|y_{line} - y_{data\,point}| = |mx_j + b - y_j|.$$

The method of Least Squares defines the best fitting line to be the line that minimizes the sum of the squares of these individual distances. Thus, we want to minimize the function

$$S(m, b) = \sum_{j=1}^{n}(mx_j + b - y_j)^2$$ where the terms x_j and y_j are considered constant.

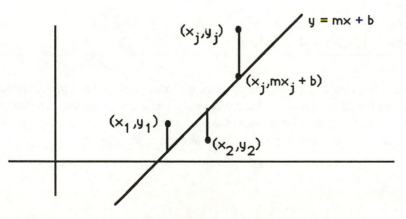

❑ **3.1.** Find the value of m and b that will minimize the function S(m,b).

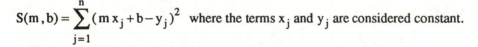

Show that $S_m = 2\left(m\sum_{j=1}^{n}x_j^2 + b\sum_{j=1}^{n}x_j - \sum_{j=1}^{n}x_j y_j \right)$ and $S_b = 2\left(m\sum_{j=1}^{n}x_j + bn - \sum_{j=1}^{n}y_j \right)$.

Since this is an arbitrary sum, *Mathematica* will not do the above derivative, that's something you will have to determine yourself. However, we can get *Mathematica* to solve for the critical points. To solve $\begin{cases} S_m = 0 \\ S_b = 0 \end{cases}$, label $\sum x_i^2$ as sumsqx, $\sum x_i$ as sumx, $\sum y_i$ as sumy and $\sum x_i y_i$ as sumxy. Thus $S_m = 2(m\,\text{sumsqx} + b\,\text{sumx} - \text{sumxy})$ and $S_b = 2(m\,\text{sumx} + b\,n - \text{sumy})$. Now use the **Solve** command to find the critical points.

Now use the **Solve** command to find the critical points.

The following table lists the relationship between semester averages and scores on the final examination for six students in a calculus class.

Semester Average	62	68	76	83	87	92
Final Examination	65	72	82	81	88	95

❏ **3.2.** Use the Method of Least Squares to find the line $y = mx + b$ that best fits the given data. To calculate m and b from a list of x and y values using the formulas established in **3.1.**, we need to calculate the sums that appear in those formulae. This can be done as specified below. State what you **Enter** and give the results.

$x = \{x_1, x_2, \text{etc} - - -\}$
$y = \{y_1, y_2, \text{etc} - - -\}$

```
sumsqx = Sum[ x[[i]]^2 , {i,1,n} ]
sumx  = Sum[ x[[i]] , {i,1,n} ]
sumy  = Sum[ y[[i]] , {i,1,n} ]
sumxy = Sum[ x[[i]] y[[i]] , {i,1,n} ]
```

❏ **3.3.** Plot the data points and approximating curve on the same graph to get a picture of how the approximating line fits the data points. This can be accomplished by using the **ListPlot** command to plot the data points, using the **Plot** command to plot the approximating curve $y = mx + b$, and then using the **Show** command to plot these two graphs on the same graph.

❏ **3.4.** Using the line of best fit, estimate the final examination grade of a student with an average of 75.

Lagrange   Multipliers

NAME(S):   **INSTRUCTOR:**
CLASS: **DATE:**

■ INTRODUCTION

The purpose of this lab is to use the method of Lagrange multiplers to solve constrained extrema problems. The method of Lagrange multipliers rests on **Lagrange's Theorem:**

Let $f(x,y)$ and $g(x,y)$ have continuous first partial derivatives and suppose f has an extremum $f(x_0,y_0)$ when (x,y) is subject to the constraint $g(x,y) = 0$. If $\nabla g(x_0,y_0) \neq 0$, then there is a real number λ such that $\nabla f(x_0,y_0) = \lambda \nabla g(x_0,y_0)$.

CAUTION. On many occasions, the resulting system of equations may be non-linear. Even though *Mathematica* is very powerful, many such systems are not easily solved.

■ NEW *MATHEMATICA* COMMANDS

There are no new *Mathematica* commands necessary for this lab.

■ METHOD OF LAGRANGE MULTIPLIERS (2 VARIABLES)

> **EXAMPLE 1.** Find the extrema of $f(x,y) = x\,y$ if (x,y) is restricted to the ellipse $4x^2 + y^2 = 4$.

Let $g(x,y) = 4x^2 + y^2 - 4$. Then the graph of $g(x,y) = 0$ is the graph of the ellipse $4x^2 + y^2 = 4$.

❑ **1.1.** Explain why $\nabla g(x,y)$ is normal to $g(x,y)=0$.

❑ **1.2.** Explain why $\nabla f(x,y)$ is normal to the level curves of $f(x,y)$.

The following diagram shows graphs of several level curves of f and a graph of the ellipse g(x,y)=0. The **method of Lagrange Multipliers** asserts that the extrema of f(x,y) subject to the constraint g(x,y)=0 occur at the points (x,y) where the level curves of f and the graph of g share a common tangent.

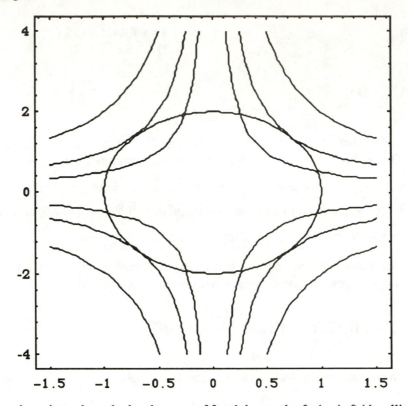

Observe that the points where the level curves of f and the graph of g(x,y)=0 (the ellipse) share a common tangent are the points where vectors perpendicular to the level curves of f and the graph of g(x,y)=0 are parallel. Since ∇f(x,y) is perpendicular to the level curves of f(x,y) and ∇g(x,y) is perpendicular to g(x,y)=0, ∇f(x,y) is parallel to ∇g(x,y) when there is a number λ satisfying ∇f(x,y)=λ∇g(x,y).

Using *Mathematica* to solve the resulting system of equations that occur when using the method of Lagrange Multipliers can make solving problems of this type much easier.

❏ **1.3.** To proceed with *Mathematica*, first carefully define f and g.

❏ **1.4.** After checking to be sure that you have defined f and g correctly, compute ∇f(x,y) and ∇g(x,y) by consecutively **Enter**ing:
gradf={D[f[x,y],x],D[f[x,y],y]}
gradg={D[g[x,y],x],D[g[x,y],y]}

182

❏ **1.5.** Find the solutions to the system of equations $\nabla f(x,y) = \lambda \nabla g(x,y)$ and $g(x,y)=0$ by **Entering:**

possibilities=Solve[{ gradf==lambda gradg, g[x,y]==0 }, {x,y,lambda}]

❏ **1.6.** To determine the maximum and minimum values, **Enter** the following command. What are the maximum and minimum values and for what points (x,y) do they occur? NOTE. This command first evaluates f(x,y) at the points (x,y) determined by lambda in the table **possibilities** and then displays the resulting output in row-column form.

{x,y,f[x,y]} /. possibilities // TableForm

EXAMPLE 2. Let $f(x,y) = 3x^2 + 2y^2 - 4y + 1$. Find the extreme values of f on the circle $x^2 + y^2 = 16$.

Let $g(x,y) = x^2 + y^2 - 16$. Then the graph of $g(x,y) = 0$ is the graph of the circle $x^2 + y^2 = 16$.

The following diagram shows graphs of several level curves of f and a graph of the circle g(x,y)=0. (The graph of the circle g(x,y)=0 is indicated in gray.) The **method of Lagrange Multipliers** asserts that the extrema of f(x,y) subject to the constraint g(x,y)=0 occur at the points (x,y) where the level curves of f and the graph of g share a common tangent.

Observe that the points where the level curves of f and the graph of g share a common tangent are the points where vectors perpendicular to the level curves of f and the graph of g(x,y)=0 are parallel. Since $\nabla f(x,y)$ is perpendicular to the level curves of f(x,y) and $\nabla g(x,y)$ is perpendicular to g(x,y)=0, $\nabla f(x,y)$ is parallel to $\nabla g(x,y)$ when there is a number λ satisfying $\nabla f(x,y) = \lambda \nabla g(x,y)$.

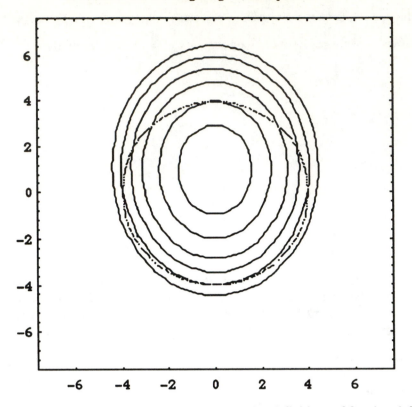

❏ **2.1.** As in the previous problem, begin by clearing all prior definitions of f and g, defining f and g, and then computing the gradient of f and g.

❏ **2.2.** What are the solutions to the system of equations $\nabla f(x,y) = \lambda \nabla g(x,y)$ and $g(x,y) = 0$?

❏ **2.3.** What are the maximum and minimum values and for what points (x,y) do they occur?

■ METHOD OF LAGRANGE MULTIPLIERS (3 VARIABLES)

EXAMPLE 3. Find the point(s) on the ellipsoid $\dfrac{x^2}{16} + \dfrac{y^2}{25} + \dfrac{z^2}{9} = 1$ closest to the point $(1, 2, -2)$.

Also find the point(s) furthermost from the point $(1, 2, -2)$.

HINT. The distance from a point (x, y, z) to the point $(1, 2, -2)$ is $\sqrt{(x-1)^2 + (y-2)^2 + (z+2)^2}$.

However, we can simplify calculations by considering the square of the distance. i.e.

$f(x, y, z) = (x-1)^2 + (y-2)^2 + (z+2)^2$. The point (x, y, z) is on the ellipsoid implies

$\dfrac{x^2}{16} + \dfrac{y^2}{25} + \dfrac{z^2}{9} = 1$. $\left(\text{i.e.} \quad \dfrac{x^2}{16} + \dfrac{y^2}{25} + \dfrac{z^2}{9} - 1 = 0. \right)$ If we let $g(x, y, z) = \dfrac{x^2}{16} + \dfrac{y^2}{25} + \dfrac{z^2}{9} - 1$, then

(x, y, z) is on the ellipsoid implies $g(x, y, z) = 0$. Thus, we want to minimize (maximize)

$f(x, y, z)$ subject to the condition $g(x, y, z) = 0$.

❏ **3.1.** Extending the method of Lagrange Multipliers to functions of three variables, find the point(s) on the ellipsoid closest to, and furthermost from, the point $(1,2,-2)$. Indicate the steps you take as well as the answers.

❏ **3.2.** Indicate the minimum and maximum distances.

EXAMPLE. 4. Let R be the region bounded by the graphs of $y = 8 - 2x^2$ and $y = x^2 - 1$.

❏ **4.1.** Plot the graphs simultaneously on the same axes.

❏ **4.2.** Find the rectangle with sides parallel to the x-axis and y-axis, and inscribed in R, having maximum area.

❏ **4.3.** Find the isosceles triangle with base parallel to the x-axis, and inscribed in R, having maximum area.

Double Integrals

NAME(S):   INSTRUCTOR:
CLASS: DATE:

▣ INTRODUCTION

If $f(x,y)$ is a non - negative integrable function on the region R, then the volume V of the solid under the surface $z = f(x,y)$ and above R is represented by $V = \iint\limits_{R} f(x,y)dA$. The area of R is represented by

$\iint\limits_{R} dA$. The purpose of this lab is to compute double iterated integrals with Mathematica.

▣ NEW *MATHEMATICA* COMMANDS

The **Integrate** command has been used before to find antiderivatives and definite integrals. We will use this command to *attempt* to evaluate double integrals in this lab. (As you recall, every Computer Algebra System (CAS) has limitations and can't handle all integrals.) In general, the command

Integrate[f[x,y],{y,c,d},{x,a,b}] calculates the iterated integral $\int_{c}^{d}\int_{a}^{b} f(x,y)\,dx\,dy$.

If you are using a version of *Mathematica* which is "pre-Version 2.0", be sure the package **IntegralTables.m** has been loaded before attempting to use the command **Integrate**. If you need to load this package, **Enter** the command **<<IntegralTables.m**

▣ EVALUATING DOUBLE INTEGRALS

The double integral $\int_{e}^{f}\int_{c}^{d} f(x,y)\,dy\,dx$ can be evaluated by applying the Integrate command as follows.

Integrate[Integrate[f[x,y],{y,c,d}, {x,e,f}]
or
Integrate[f[x,y],{x,e,f},{y,c,d}]

■ DOUBLE INTEGRALS

EXAMPLE 1. Use *Mathematica* to compute $\int_0^1 \int_0^3 f(x,y)\,dy\,dx$ if $f(x,y) = x\sqrt{x^2 + y}$.

❑ **1.1.** To compute $\int_0^1 \int_0^3 f(x,y)\,dy\,dx$ first define f and then evaluate the integral by **Entering** the following commands:

f[x_,y_]:=x Sqrt[x^2+y]
Integrate[f[x,y],{x,0,1},{y,0,3}]

What is the region R in this problem?

EXAMPLE 2. Use *Mathematica* to compute $\int_0^1 \int_x^1 f(x,y)\,dy\,dx$ if $f(x,y) = x\,e^{y}$.

❑ **2.1.** Enter the following three commands and interpret each result.

Clear[f]
f[x_,y_:=x Exp[y^3]

Integrate[f[x,y],{x,0,1},{y,x,1}]

What does this integral represent? Be specific.

Let $R = \{(x,y) : 0 \le x \le 1,\ x \le y \le 1\}$. Then, $\int_0^1 \int_x^1 f(x,y)\,dy\,dx = \int_R f(x,y)\,dy\,dx$. Moreover,

$R = \{(x,y) : 0 \le y \le 1,\ 0 \le x \le y\}$ so $\int_R f(x,y)\,dy\,dx = \int_0^1 \int_0^y f(x,y)\,dx\,dy$. Sketch the region R.

❑ **2.2.** Compute $\int_0^1 \int_x^1 f(x,y)\,dy\,dx$ by using *Mathematica* to evaluate $\int_0^1 \int_0^y f(x,y)\,dx\,dy$.

EXAMPLE 3. Find values of a and b so that both $\int_0^{\frac{\pi}{2}} \int_0^{\frac{\pi}{2}} (b\,y\cos(x) + a\,x\sin(y))\,dy\,dx = 1$

and $\int_0^{\pi} \int_0^{\pi} (b\,y\cos(x) + a\,x\sin(y))\,dx\,dy = 1$.

❑ **3.1.** First clear all prior definitions of a, b, and f and then define

$f(x,y) = b\,y\cos(x) + a\,x\sin(y)$ by **Enter**ing:

Clear[a,b,f]
f[x_,y_]:=b y Cos[x]+a x Sin[y]   (* Don't forget to include the spaces! Why? *)

❑ **3.2.** Evaluate $\int_0^{\frac{\pi}{2}} \int_0^{\frac{\pi}{2}} f(x,y)\,dy\,dx$ and name the result **lhsone**. Indicate the command(s) you used as well as the result.

❑ **3.3.** Evaluate $\int_0^{\pi} \int_0^{\pi} f(x,y)\,dy\,dx$ and name the result **lhstwo**. Indicate the command(s) you used as well as the result.

❑ **3.4.** Solve the system of equations **lhsone==1**, **lhstwo==1**. What are the values of a and b that solve the problem?

EXAMPLE 4. Let R be the region bounded between the graphs of $x^2 + y^2 = 9$ and $x^2 + (y-2)^2 = 4$.

❑ **4.1.** Display both graphs on the same axes and find the exact points of intersection.

❑ **4.2.** Give an exact description of the region R. (i.e. {(x,y): ....})

❑ **4.3.** Represent the area in terms of an integral of the form $\iint_R dA = \int_a^b \int_{f(x)}^{g(x)} dy\, dx$.

❑ **4.4.** Calculate both the exact value and a numerical approximation of the area of R.

Triple Integrals

NAME(S):
CLASS:

INSTRUCTOR:
DATE:

■ INTRODUCTION

In this assignment we will look at how to **evaluate triple integrals** and then we will look at some situations whose solution involve evaluating triple integrals. One of the obvious applications of triple integrals is to enlarge the class of solids for which we can find the **volume**. With the aid of the triple integral we are able to obtain the volume of any solid which is formed by intersecting a finite number of curved surfaces in 3-space. In the section on finding volumes, we attempt to visualize the "skeleton" of the solid by determining the boundary curves that make up the sides of the solid.

■ NEW *MATHEMATICA* COMMANDS

The **Integrate** command has been used before to find antiderivatives and definite integrals. We will use this command to *attempt* to evaluate triple integrals in this lab. (As you recall, every Computer Algebra System (CAS) has limitations and can't handle all integrals.) The commands **Line** and **Graphics3D** are used in one section and are explained at that time.

If you are using a version of *Mathematica* which is "pre-Version 2.0", be sure the package **IntegralTables.m** has been loaded before attempting to use the command **Integrate**. If you need to load the integral tables, **Enter** the command **<<IntegralTables.m**

■ EVALUATING TRIPLE INTEGRALS

The triple integral $\int_e^f \int_c^d \int_a^b f(x,y,z)\,dz\,dy\,dx$ can be evaluated by applying the Integrate command as follows.

Integrate[Integrate[Integrate[f[x,y,z],{z,a,b}],{y,c,d}],{x,e,f}] or
Integrate[Integrate[f[x,y,z],{y,c,d},{z,a,b}], {x,e,f}] or
Integrate[f[x,y,z],{x,e,f},{y,c,d},{z,a,b}]

EXAMPLE 1. Evaluate $\int_e^f \int_c^d \int_a^b f(x,y,z)\,dz\,dy\,dx$.

❑ **1.1.** Evaluate $\int_{-2}^3 \int_0^3 \int_1^2 (z^2 x + x y)\,dz\,dy\,dx$.

Integrate[z^2 x + x y ,{x,-2,3},{y,0,3},{z,1,2}]

❏ **1.2.** Evaluate the integral in **1.1.** by the other possible variations in using the Integrate command. Indicate each command as well as its corresponding answer.

❏ **1.3.** Evaluate $\displaystyle\int_{\frac{\pi}{6}}^{\frac{\pi}{3}} \int_{1}^{3} \int_{0}^{2} \sin(x^2 z)\,(x^2 + y^2)\,dz\,dy\,dx$.

■ USING TRIPLE INTEGRALS TO EVALUATE THE VOLUME OF SOLIDS GENERATED BY INTERSECTING CURVED SURFACES IN 3-SPACE

The following example shows how one can build the boundary curves that make up the outside surface of the object. This will greatly enhance our ability to write down the triple integral that represents the volume of the solid.

EXAMPLE 2. Find the volume under the surface $z = 3x^3 + 3x^2y$ and above the rectangle $R = \{(x,y)\,|\,1 \le x \le 3,\ 0 \le y \le 2\}$.

❏ **2.1.** First, plot the surface by **Enter**ing the following commands.
Clear[f]
f[x_,y_]:=3 x^3 + 3 x^2 y
Plot3D[f[x,y],{x,1,3},{y,0,2}]

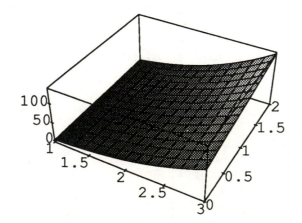

From this graph, it's fairly obvious that the solid is bounded above by z=f(x,y), below by the rectangle R in the xy-plane, and on the sides by exterior surfaces which we will label as leftface, rightface, frontface, and backface.

The **leftface** is the trapezoidal region whose corner points are P0={1,0,0}, P1={1,2,0}, P2={1,2,f(1,2)}, and P3={1,0,f(1,0)}.

The **rightface** consist of the trapezoidal region joining Q0={3,0,0}, Q1={3,2,0}, Q2={3,2,f(3,2)}, and Q3={3,0,f(3,0)}.

The **frontface** consists of the horizontal line segment from P0 to Q0, the vertical line segment from Q0 to Q3, the curve **curve1** from Q3 to P3 and the vertical line segment from P3 to P0.

The **backface** consists of the horizontal line segment fromP1 to Q1, the vertical line segment from Q1 to Q2, the curve **curve2** from Q2 to P2 and the vertical line segment from P2 to P1.

❑ **2.2.** Sketch the figure showing the four exterior surfaces. Label the surfaces, points, and curves indicated above.

To define the trapezoidal regions we can create its boundary points and then join these points with straight line segments using the **Line** command.

❑ **2.3.** Define **leftface** by **Enter**ing the following commands.
P0={1,0,0}
P1={1,2,0}
P2={1,2,f[1,2]}
P3={1,0,f[1,0]}
leftface=Line[{P0,P1,P2,P3,P0}]

NOTE. Since the **Line** command draws a sequence of line segments connecting the specified points, if we want to draw an enclosed figure we must begin and end with the same point. This object can be drawn using **Show[Graphics3D[frontface],BoxRatios->{.5,.5,.4}]**
(The command **Graphics3D** must be included in order to instruct *Mathematica* that this is a line segment in 3-space instead of 2-space.)

❑ **2.4.** Define **rightface** by **Enter**ing the following commands.

Q0={3,0,0}
Q1={3,2,0}
Q2={3,2,f[3,2]}
Q3={3,0,f[3,0]}
 rightface=Line[{Q0,Q1,Q2,Q3,Q0}]

There are two different approaches we can use for plotting the curved surface. One approach involves finding a parametric equation for the curve and then plotting this curve. For geometric solids, we can usually view a boundary curve as the lift by the function f(x,y) of a specified curve in the xy-plane. This is similar to the treatment we employed in LAB # 30 on directional derivatives. Another approach involves constructing the curved boundary by plotting a finite number of points (20 or so) along this curve and then connect these points with a straight line. This is the approach we will take for this example. We need to take care of one technicality, namely: will a finite number of line segments generate a reasonably accurate picture of the curved surface? The answer to this question is yes provided we have a curve that is continuous and reasonably smooth. That is true for the boundary curves of this solid. To generate a uniformily distributed set of points on the curve, we will use the **Table** command and then use the **Line** command to join these together along with the other corner points of the frontface structure. Be sure you understand what the following commands are doing.

❑ **2.5.** Define **frontface** by **Enter**ing the following commands.

curve1=Table[{x,0,f[x,0]},{x,1,3,.1}]
frontface=Line[Flatten[{{P0},curve1,{Q0,P0}},1]]

Notice also, since P3={1,0,f[1,0]} is the first point on the list **curve1** and Q3={3,0,f[3,0]} is the last point on the list **curve1**, we have properly ordered the points that are to be connected with line segments. The **Flatten[list,1]** command is used to remove the first layer of braces in the list created with the Table command. In a similar fashion we can create the backface structure.

❑ **2.6.** Define **backface** by **Enter**ing the following commands.

curve2=Table[{x,2,f[x,2]},{x,1,3,.1}]
backface=Line[Flatten[{{P1},curve2,{Q1,P1}},1]]

❑ **2.7.** Plot all four faces by executing the command

Show[Graphics3D[{PointSize[.02],frontface,backface,leftface,rightface,
      Point[P0],Point[P3],Point[Q3],Point[Q0]}],BoxRatios->{1,1,.7},Boxed->False]

We have included the four points {P0,P3,Q3,Q0} in order to orient the solid in our picture. The option **PointSize[.02]** is used to enlarge the points so they will be more visible. Does this resemble your sketch in 2.2.?

You may want to go back over the previous commands in order to obtain a clearer understanding of how you can:
(i)   draw a polygonal shaped figure in 3-space,
(ii) draw a curved shape in 3-space,
(iii) merge two or more boundaries,
(iv) plot the image of two or more surfaces on the same graph, and
(v)   include several points on a graph in 3-space so as to orient the position of a given object.

❏ **2.8.** Write down the triple integral(s) that represents the volume.

❏ **2.9.** Find the volume by evaluating your integral(s) above.

EXAMPLE 3. Find the volume of the solid in the first octant that is enclosed by the planes x=0, z=0, x=5, z = y, and z =-2y + 6.

First, define the functions f1 and f2 by **Enter**ing the following commands.
Clear[f1,f2]
f1[x_,y_]:=y
f2[x_,y_]:=-2y+6

❏ **3.1.** Plot the xy-plane by **Enter**ing the command **xyplane=Plot3D[0,{x,0,5},{y,0,5}]**

❏ **3.2.** Plot the plane z=f1(x,y) by **Enter**ing the command
gf1=Plot3D[f1[x,y],{x,0,5},{y,0,5}]

❏ **3.3.** Plot the plane z=f2(x,y) by **Enter**ing the command
gf2=Plot3D[f2[x,y],{x,0,5},{y,0,5}]

❏ **3.4.** To merge the 3 planes together, one can enter the command
Show[Graphics3D[gf1],Graphics3D[gf2],Graphics3D[xyplane],
ViewPoint->{5,-5,5}]
Since this is VERY memory hungry and takes lots of time, the result is shown below.

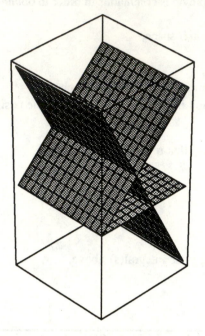

❏ **3.5.** This solid is formed by the intersection of five planes. What are the equations of these five planes?

Proceeding as we did before, we can plot the outline of this solid. It is given below. The bold points are the corner points where x=0.

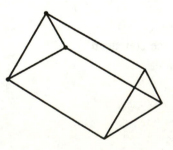

❏ **3.6.** Write down the triple integral(s) that represents the volume and find the volume by evaluating your integral(s).

More on Volumes
and
Triple Integrals

NAME(S):   INSTRUCTOR:

CLASS: DATE:

▣ INTRODUCTION

In this lab we will look further at calculating volumes by triple integrals.

▣ NEW *MATHEMATICA* COMMANDS

The **Integrate** command is the one most used in this lab and it is certainly not new to you. Be sure to remember the Integrate command has limitations. As you recall, every Computer Algebra System (CAS) has limitations and can't handle all integrals.

If you are using a version of *Mathematica* which is "pre-Version 2.0", be sure the package **IntegralTables.m** has been loaded before attempting to use the command **Integrate**. If you need to load the integral tables, **Enter** the command **<<IntegralTables.m**

▣ EVALUATING TRIPLE INTEGRALS

The triple integral $\int_e^f \int_c^d \int_a^b f(x,y,z)\, dz\, dy\, dx$ can be evaluated by applying the Integrate command as follows.

Integrate[Integrate[Integrate[f[x,y,z],{z,a,b}],{y,c,d}],{x,e,f}] or
Integrate[Integrate[f[x,y,z],{y,c,d},{z,a,b}], {x,e,f}] or
Integrate[f[x,y,z],{x,e,f},{y,c,d},{z,a,b}]

▣ VOLUME

Let f(x,y) be a non-negative integrable function on a region R. Then the volume V of the solid that lies

above R and under the surface z = f(x,y) is $V = \iint_R f(x,y)\, dA = \iint_R \int_0^{f(x,y)} 1\, dz\, dA$.

EXAMPLE 1. Find the volume of the solid bounded by the surface $z = y\sqrt{x + y^2}$, the planes $x = 0, y = 0$, and $x + y = 1$.

❏ **1.1.** Let R denote the region in the xy-plane bounded by x=0, y=0, and x+y=1. Explain why R can be described by R={(x,y): $0 \leq x \leq 1, 0 \leq y \leq 1$-x}.

❏ **1.2.** Explain why the volume V is given by $V = \iint\limits_R y\sqrt{x + y^2}\, dA = \int_0^1 \int_0^{1-x} y\sqrt{x + y^2}\, dy\, dx$.

❏ **1.3.** Calculate the volume V.

EXAMPLE 2. Use a triple integral to find the volume of the tetrahedron T bounded by the planes 2x+4y+z=2, x=3y, x=0, and z=0.

❏ **2.1.** Explain why the projection of T onto the xy − plane is the region bounded by the graphs of $x = 0, y = \dfrac{x}{3}$, and $y = \dfrac{1-x}{2}$.

❏ **2.2.** Graph $y = \dfrac{x}{3}$ and $y = \dfrac{1-x}{2}$ at the same time on the interval [0,1] and locate the intersection point.

❏ **2.3.** Explain why the volume of T is given by the triple integral

$\int_0^{\frac{3}{5}} \int_{\frac{x}{3}}^{\frac{1-x}{2}} \int_0^{2-2x-4y} 1\, dz\, dy\, dx$.

❏ **2.4.** Calculate the volume of T.

■ USING TRIPLE INTEGRALS TO EVALUATE THE VOLUME OF SOLIDS GENERATED BY INTERSECTING CURVED SURFACES IN 3-SPACE

> **EXAMPLE 3.** Find the volume of the solid enclosed between the two paraboloids $z = 8 - x^2 - y^2$ and $z = 3x^2 + y^2$.

Let **surface1** denote the lower paraboloid and **surface2** denote the upper paraboloid.

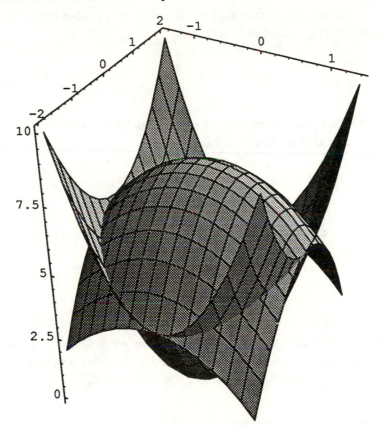

❏ **3.1.** Find the equation of the curve generated by the intersection of surface1 and surface2. Generate the plot of curveinter = surface1 ∩ surface2.
HINT. The projection of this curve onto the xy-plane is an ellipse. Find the equation of this ellipse.

❏ **3.2.** Find the equation of the curve generated by intersecting the xz-plane and surface2. Generate the plot of Curvexz2 = xz-plane ∩ surface 2.

❏ **3.3.** Find the equation of the curve generated by intersecting the yz-plane and surface 2. Generate the plot of Curveyz2 = yz-plane ∩ surface 2.

❏ **3.4.** Find the equation of the curve generated by intersecting the xz-plane and surface1. Generate the plot of Curvexz1=xz-plane ∩ surface 1.

❏ **3.5.** Find the equation of the curve generated by intersecting the yz-plane and surface1. Generate the plot of Curveyz1 = yz-plane ∩ surface 1.

❏ **3.6.** Merge the curves obtained from 3.1-3.5 into a common graph. Plot this picture.
If you have a printing capability then print this graph using the **Print Selection** option in the **File** menu. Otherwise, make sure this graph is included on your disk.

❏ **3.7.** Write down the triple integral that represents the volume.

❏ **3.8.** Find the volume by evaluating your integral above.

Surface Area & Other Applications of Multiple Integrals

NAME(S):   INSTRUCTOR:
CLASS: DATE:

▣ INTRODUCTION

We have already used multiple integrals to represent and calculate volumes. The purpose of this lab is to examine other applications of multiple integrals. In particular, we will discuss calculating surface area and mass. We will also examine the Monte Carlo method of estimating certain integrals, areas, and volumes.

▣ NEW *MATHEMATICA* COMMANDS

There are several new user defined commands that will be used with the Monte Carlo method of estimating integrals. These commands have been created for use in this lab and are available from your instructor.

1. ptinregion[x,y,z]
   This command will return the value TRUE or FALSE depending upon whether or not the point (x,y,z) is in the region Q. This command is used to actually define the region Q. (See Example 3.) Since the user defines the particular region under consideration with this command, it CANNOT be defined for you.

2. generatepts[s,n,a,b]
   This command is used to create a set **s** of **n** random points (x,y,z) in the region Q. The values x, y, and z are generated in the interval [a,b]. Since this generates points in the region Q, the command ptinregion[x,y,z] **MUST** have already been defined and **Enter**ed!!! Note that since x,y, and z are in [a,b], **a** and **b** must be chosen so that Q is a subset of [a,b]×[a,b]×[a,b].

3. sumf[f,s]
   This command finds the sum of the f(x,y,z)'s such that (x,y,z) is in the set **s**.

▣ SURFACE AREA

Let z=f(x,y) have continuous first partial derivatives on a closed region R in the xy-plane. Then the area of the surface z=f(x,y) for (x,y) in R is given by the double integral

Surface Area $= \iint\limits_{R} \sqrt{\left(\dfrac{\partial z}{\partial x}\right)^2 + \left(\dfrac{\partial z}{\partial y}\right)^2 + 1}\ \ dA = \iint\limits_{R} \sqrt{\left(f_x(x,y)\right)^2 + \left(f_y(x,y)\right)^2 + 1}\ \ dA.$

EXAMPLE 1. Find the surface area of the portion of the cone $z = 2\sqrt{x^2 + y^2}$ inside the cylinder $x^2 - 2x + y^2 - 2y = 2$.

Completing the square on the cylinder $x^2 - 2x + y^2 - 2y = 2$ yields $(x-1)^2 + (y-1)^2 = 4$ so that $y = 1 \pm \sqrt{4 - (x-1)^2}$.

Begin by carefully defining $f(x, y) = 2\sqrt{x^2 + y^2}$.

❏ **1.1.** Compute $\dfrac{\partial f}{\partial x}$ and naming the result **fx** and $\dfrac{\partial f}{\partial y}$ and naming the result **fy** by **Enter**ing the following commands. Indicate the result of each.

fx=D[f[x,y],x]

fy=D[f[x,y],y]

❏ **1.2.** Compute and simplify $\sqrt{\left(\dfrac{\partial f}{\partial x}\right)^2 + \left(\dfrac{\partial f}{\partial y}\right)^2 + 1}$ by **Enter**ing each of the following commands. Indicate the result of each.

Sqrt[fx^2+fy^2+1]

integrand=Sqrt[Together[fx^2+fy^2+1]]

❏ **1.3.** Explain why the desired surface area is given by the integral $\displaystyle\int_{-1}^{3} \int_{1-\sqrt{4-(x-1)^2}}^{1+\sqrt{4-(x-1)^2}} \sqrt{5}\ \ dy\,dx$ and evaluate using *Mathematica*.

202

❏ **1.4.** By hand, the integral $\int_{-1}^{3} \int_{1-\sqrt{4-(x-1)^2}}^{1+\sqrt{4-(x-1)^2}} \sqrt{5} \; dy \, dx$ is easier to evaluate if converted to polar coordinates. Convert the integral to polar coordinates.

▣ TRIPLE INTEGRALS THAT ARISE FROM NONGEOMETRIC SETTINGS

EXAMPLE 2. The mass m of a solid Q can be determined by evaluating the integral

$$m = \iiint_Q \delta(x,y,z) \, dV, \text{ where } \delta(x,y,z) \text{ is the density at the point .}$$

Find the mass of the solid Q bounded by the surfaces whose equations are $x^2 + z = 4$ and $y + z = 4$, if $\delta(x,y,z) = z(x^2 + y)$. Indicate your steps as well as your results.

EXAMPLE 3. Monte Carlo Method For Estimating Integrals

The average value of f(x,y,z) over a region Q of volume V is defined by

$$f_{Average} = \frac{1}{V} \iiint_Q f(x,y,z) \, dV.$$

If points $(x_1,y_1,z_1), ..., (x_n,y_n,z_n)$ are uniformly distributed throughout Q then

$$f_{Average} \approx \frac{1}{n} \sum_{k=1}^{n} f(x_k,y_k,z_k).$$

Thus

$$\iiint_Q f(x,y,z) \, dV = V \bullet f_{Average} \approx \frac{V}{n} \sum_{k=1}^{n} f(x_k,y_k,z_k).$$

Note that when estimating $\int_{e}^{f} \int_{c}^{d} \int_{a}^{b} f(x,y,z) \, dz \, dy \, dx$, the region is a parallelopiped with volume $V = (b-a)(d-c)(f-e)$.

203

This approach is called the **Monte Carlo method** of estimating triple integrals. (This approach can also be used for single and double integrals.) It is not generally competitive with good algorithms, but it can be attractive in higher dimensions with certain types of functions. It should be noted that n has to be fairly large in order to get a good estimate. It should also be noted that the volume of the region must be known in order to estimate the integral.

One can also estimate volumes and areas by Monte Carlo techniques. This is particularly useful when trying to find the volume, or area, of an irregularly shaped region. (See Exercise 5.)

❑ **3.1.** Use the Monte Carlo method with n = 100 to approximate the mass of the solid with shape Q where Q is the region in the first octant bounded by the portion of the sphere

$x^2 + y^2 + z^2 = 1$ with density function $f(x, y, z) = x + y + z$.

First, use ptinregion to define the region Q described above by **Enter**ing the commands
Clear[ptinregion]
ptinregion[{x_,y_,z_}]:=x>=0 && y>=0 && z>= 0 && x^2+y^2+z^2 <=1

(The symbol **&&** denotes the **logical and** operation. <u>REMEMBER</u>. Be sure to **Enter** the above command before executing the command generatepts!)

Now, generate 100 points within the sphere by **Enter**ing the following commands.
Clear[s]
generatepts[s,100,-1,1]

After defining the function f, **Enter** the command **sumf[f,s]** to evaluate f at each of these points and sum the results. State the result of **Enter**ing the command and then give the estimate for the mass.

❑ **3.2.** Because f(x,y,z) is a relatively simple function, we can perform the integration and thus find the true mass. What is the correct value of the triple integral used to define the mass? Indicate the steps taken as well as your answer.

❑ **3.3.** How many points would have to be generated in order for the difference between the actual values of the integral and the value returned from the Monto Carlo method to be less than .05? Indicate the steps taken as well as your answer.

Cylindrical and Spherical Coordinates

NAME(S):   INSTRUCTOR:
CLASS: DATE:

■ INTRODUCTION

When working with equations that describe certain shapes, changing coordinate systems can sometimes reduce the given equation to a much simpler one. For example, when working with equations that describe cylinders or spheres, converting to cylindrical or spherical coordinates may reduce the equation to a much simpler one and consequently the problem will be easier. The purpose of this lab is to introduce the cylindrical and spherical coordinate systems and then compute integrals using these coordinates.

■ NEW *MATHEMATICA* COMMANDS

This lab uses several commands, **Coordinates** and **VolumeElement,** defined in the package **VectorAnalysis.m** which is contained in the **Calculus** folder (or directory). One way to load the package is to **Enter** the command <<Calculus`VectorAnalysis. In addition, be sure the package **IntegralTables.m** has been loaded prior to computing any exact definite integrals.

1. **Coordinates[system]** gives a list of the names of coordinates in the coordinate system **system**.

2. **VolumeElement[system]** gives the volume element for the coordinate system **system**.

■ CYLINDRICAL COORDINATES

Each point (x,y,z) in space can be represented in the **cylindrical coordinate system** by the

coordinates (r, θ, z), where $\begin{cases} x = r\cos(\theta) & r^2 = x^2 + y^2 \\ y = r\sin(\theta) & \tan(\theta) = \dfrac{y}{x} \end{cases}$.

EXAMPLE 1. Describe the graph of the equation in the cylindrical coordinate system r=csc(θ).

❏ **1.1.** Note that $r = \csc(\theta) = \dfrac{1}{\sin(\theta)}$ is equivalent to $r\sin(\theta) = 1$. Replace $r\sin(\theta)$ by y

in the equation by **Entering:**

r Sin[theta]==1 /. r Sin[theta]->y

❏ **1.2.** Use **1.1** to describe the graph of the equation.

EXAMPLE 2. Find the volume of the solid bounded by the cylinder $x^2 + y^2 = 2$, the paraboloid $z = x^2 + y^2$, and the xy − plane.

❏ **2.1.** Let R denote the solid. Then, the volume of the solid is given by the triple integral

$\displaystyle\iiint_R 1\,dV$. Calculate the volume element for cylindrical coordinates by **Entering:**

VolumeElement[Cylindrical]

❏ **2.2.** Convert the equations $x^2 + y^2 = 2$ and $z = x^2 + y^2$ to cylindrical coordinates by **Entering:**

tocylindricalrule={x->r Cos[theta],y->r Sin[theta],z->z}

x^2+y^2==2 /. tocylindricalrule // TrigExpand

z==x^2+y^2 /. tocylindricalrule // TrigExpand

❑ **2.3.** Explain why $\iiint\limits_{R} 1\,dV = \int_{-\sqrt{2}}^{\sqrt{2}} \int_{-\sqrt{2-x^2}}^{\sqrt{2-x^2}} \int_{0}^{x^2+y^2} 1\,dz\,dy\,dx$

$$= \int_{0}^{2\pi} \int_{0}^{\sqrt{2}} \int_{0}^{r^2} r\,dz\,dr\,d\theta.$$

❑ **2.4.** Use *Mathematica* to evaluate each of the above integrals. Explain the steps necessary to evaluate each integral by hand and explain why the second integral is easier to evaluate by hand than the first.

■ **SPHERICAL COORDINATES**

Each point (x,y,z) in space can be represented in the **spherical coordinate system** by the

coordinates (ρ, θ, ϕ), where
$$\begin{aligned} x &= \rho \sin(\phi)\cos(\theta) \\ y &= \rho \sin(\phi)\sin(\theta) \\ z &= \rho \cos(\phi) \\ \rho &= \sqrt{x^2 + y^2 + z^2} \end{aligned}$$

EXAMPLE 3. Describe the graph of the equation in the spherical coordinate system $\rho = 4\cos(\phi)\sin(\theta)$.

❑ **3.1.** Replace ρ by $\sqrt{x^2 + y^2 + z^2}$, $\cos(\phi)$ by $\dfrac{z}{\rho}$, and $\sin(\phi)\sin(\theta)$ by $\dfrac{y}{\rho}$

by **Entering:**

```
stepone=rho^2==4 Cos[phi] Sin[phi] Sin[theta] //.
        { rho->Sqrt[x^2+y^2+z^2],Cos[phi]->z/rho, Sin[phi]  Sin[theta]->y/rho}
```

❏ **3.2.** Use **3.1** to describe the graph of the equation.

EXAMPLE 4. $\displaystyle\int_{-4}^{4} \int_{-\sqrt{16-x^2}}^{\sqrt{16-x^2}} \int_{0}^{\sqrt{16-x^2-y^2}} z^2 \sqrt{x^2 + y^2 + z^2}\ dz\ dy\ dx$

❏ **4.1.** Graph the solid region R of integration. As usual, state the commands you use as well as the result.

❏ **4.2.** Convert the above integral into an integral in terms of spherical coordinates and evaluate this new integral.

❏ **4.3.** State the command to evaluate the integral in its original form. Can *Mathematica* handle it?

▣ GRAPHING IN CYLINDRICAL COORDINATES

In Version 2.0 (or later) of *Mathematica*, the graphics capabilities include graphing in cylindrical coordinates with the **CylindricalPlot3D** command. The general form of this command is **CylindricalPlot3D[f[r,theta], {r,rmin,rmax}, {theta,thetamin,thetamax}]** Before this command can be executed, you must load the appropriate graphics package by executing the command **<<Graphics`ParametricPlot3D`**

By **Entering** the command **CylindricalPlot3D[r^2 Cos[2 theta],{r,0,1},{theta,0,2Pi}]** the following graph was obtained.

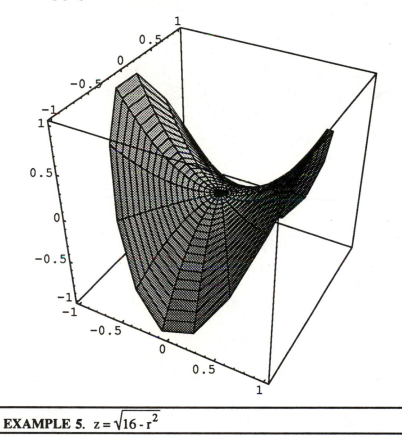

EXAMPLE 5. $z = \sqrt{16 - r^2}$

❑ **5.1.** Find appropriate intervals for r and θ and use **CylindricalPlot3D** to graph this function.

EXAMPLE 6. z=r

❑ **6.1.** Find appropriate intervals for r and θ and use **CylindricalPlot3D** to graph this function.

❑ **6.2.** Convert z=r into rectangular coordinates.

EXAMPLE 7. z=r cos(θ)

❑ **7.1.** Find appropriate intervals for r and θ and use **CylindricalPlot3D** to graph this function.

Exercises

LAB # 1 — *Introduction to Computing with Mathematica*

1. Use the command **?** to describe the syntax and action of each of the commands **Plot, Solve, NRoots, /., Together, Expand, Table, TableForm, Map,** and **N.**
   Use these commands to compute each of the following:

   (a) Graph the polynomial $4 + x - 4x^2 - 5x^3 - 10x^4 + 2x^5$ on the interval $[-1, 5.6]$.

   (b) *Mathematica* denotes the trigonometric functions $\sin(x)$, $\cos(x)$, and $\tan(x)$ by **Sin[x], Cos[x],** and **Tan[x].** *Mathematica* denotes the constant π by **Pi.**
   Use the **Plot** command to simultaneously graph the above three trigonometric functions for $-2\pi \le x \le 2\pi$, in different shades of gray.

   (c) Find the exact values of x that satisfy each equation:

   (i) $-2 + 15x + x^2 = 0$;

   (ii) $\dfrac{9x}{-3x - 9} = 7x$; and

   (iii) $\dfrac{3 + 2x + 7x^2}{-3 - 5x + 2x^2} = 4$.

   (d) Use *Mathematica* to define $k(x) = 4 + x - 4x^2 - 5x^3 - 10x^4 + 2x^5$. Approximate all real solutions to the equations $k(x) = 0$ and $k(x) = x$. How many real solutions does each equation have?

   (e) Evaluate each expression for the indicated value of x:

   (i) $\dfrac{273}{50}x$ when x has value $\dfrac{2793}{500}$;

   (ii) $\dfrac{8 - 6x^2 - 3x^3}{\sqrt[3]{10 - 2x - 6x^2 + 8x^3}}$ when x has value -10×4^{-5}.

   (f) Write each expression as a fraction with a single denominator.

   (i) $\dfrac{3}{-6 + 9x} + \dfrac{9}{5 - x} + 1$; and

   (ii) $\dfrac{-2 + 5x}{-2 - 4x - 9x^2} + \dfrac{-4 + 7x}{-2 + 8x - 3x^2} + \dfrac{-4 + 9x}{9 + 5x + 10x^2}$.

   (g) Compute the indicated multiplication:

   (i) $(2 - 5x)^5$;

   (ii) $\left(4 + 9x + 4x^2\right)^{10}$; and

   (iii) $\left(-3 - x + 10x^2\right)^3 (2 + 3x)^2$.

(h) *Mathematica* denotes the trigonometric functions sin(x), cos(x), and tan(x) by **Sin[x]**, **Cos[x]**, and **Tan[x]**. *Mathematica* denotes the constant π by **Pi**. Use the **Table** and **Map** commands to create a table of the principal values of the three trigonometric functions.

(i) Compute a 25-digit approximation of π.

2. As discussed above, the command **Solve** can sometimes be used to find exact solutions of certain equations; the command **NRoots** can be used to approximate roots of polynomial equations. Since many equations are neither solvable nor polynomials, *Mathematica* has another command, **FindRoot**, which can be used to approximate roots of some equations.

For example, if $f(x) = x \cos(x) - \sin(x)$, define f and graph f(x) for $0 \le x \le 5\pi$ by carefully typing and entering the following commands

Clear[f]
f[x_]=x Cos[x]-Sin[x]

Plot[f[x],{x,0,5Pi}]

Notice that on the interval $(0,5\pi)$, f has four zeros. Notice that the first zero occurs near 5. To use **FindRoot** to approximate the value of x near 5 that results in f(x)=0, carefully type and **Enter**:

FindRoot[f[x]==0,{x,5}]

Hence, the first value of x that results in f(x)=0 is approximately 4.49341. Similarly, observe that the second zero occurs near 7.5. To use **FindRoot** to approximate the value of x near 7.5 that results in f(x)=0, carefully type and enter:

FindRoot[f[x]==0,{x,7.5}]

The result means that the second value of x that results in f(x)=0 is approximately 7.72525. In general, the command **FindRoot** works well as long as (i) the graph of the function has both no breaks and no sharp peaks; and (ii) your "guess" is reasonably close to the actual value of the root.

(a) Use **FindRoot** to approximate the third and fourth zeros of f.

(b) Use **FindRoot** to approximate the first four solutions of the equations

   (i) x cos(x)-sin(x)=3; and
   (ii) x cos(x)-sin(x)=-2.

LAB # 2 — *Functions and Limits of Functions*

1. Suppose P dollars is invested at an annual interest rate r * 100%. If the accumulated interest is credited to the account at the end of each year, then the interest is said to be compounded annually; if it is credited at the end of each six-month period then it is said to be compounded semi-annually; and if it is credited at the end of each three-month period, then it is said to be compounded quarterly.

❑ EXERCISES ❑

The more frequently the interest is compounded, the better it is for the investor since more of the interest is itself earning interest. Lets look at some examples to see why this is true.

It can be shown that if interest is compounded n times a year at equally spaced intervals, then the value A of the investment after t years is:

$$A = P\left(1 + \frac{r}{n}\right)^{nt}.$$

Suppose the initial investment was P = \$1000.00 at an interest rate of 10% (i.e. r=0.1)

(a) Create a table showing the amount of money earned at the end of 10, 20 and 30 years if the interest is compounded
  1) semi-annually;
  2) quarterly;
  3) daily;
  4) hourly;
  5) each second.

HINT. Define a function A[t,n] = P (1 + r/n)^(n t) and declare P=1000 and r=.1. Use the command **tbl =Table[** {A[[t,semi] , A[t,quart] , A[t,daily] , A[t,hr] , A[t,sec] },{t,10,30,10}] and then use the command **TableForm[tbl]** to display these values in tabular form.

Notice that while the amount does seem to increase as the number of interest periods are increased, the increase is diminishing. Thus while the amount earned is noticeably more if the interest is compounded hourly instead of semi-annually; there is not as much of a difference between the amount earned when the interest is compounded hourly or compounded each second. This suggests that there is a limiting value which is attained as the number of periods is allowed to increase, that is,

$$\lim_{n \to \infty} P\left(1 + \frac{r}{n}\right)^{nt} = P \lim_{n \to \infty} \left(1 + \frac{r}{n}\right)^{nt}$$

must be a fixed number

(b) What is $\lim_{n \to \infty} \left(1 + \frac{r}{n}\right)^{nt}$?

HINTS. (1) Replace n with r / x; (2) Replace $\lim_{n \to \infty} *$ with $\lim_{x \to ?} *$; and

(3) Use the fact that $\lim_{x \to a} f[x]^u = \left(\lim_{x \to a} f[x]\right)^u$.

Compare to the function discussed in EXAMPLE 2.

2. Let $f(x) = \dfrac{\sin(x)}{x}$ and a = 0.

(a) Determine the range of fluctuation of f over each of the following deleted neighborhoods of a by doing the following:

 1. Plot f(x) over a symmetric interval about x=0 of width .2
 2. Use the graph of y = f(x) to determine the largest value of f over this interval.
 3. Use the graph of f to determine the smallest value of f over this interval,
 4. What is the difference between the largest value and the smallest value.
    (i)   a symmetric interval about x=0 of length .2
    (ii) a symmetric interval about x=0 of length .02.
    (iii) a symmetric interval about x=0 of length .0002.

(b) Determine the range of fluctuation of f over each of the following deleted neighborhoods by doing the following.

 1. Generate 100 random values of x that lie in an interval of width .02 about x=0.
 2. Find the largest value of f over this set of points.
 3. Find the smallest value of f over this set of points.
    (i)   a symmetric interval about x=2 of length .2
    (ii) a symmetric interval about x=2 of length .02.
    (iii) a symmetric interval about x=2 of length .0002.

(c) Looking at the above results, would you say that the limit of f(x) at x=0 exists?
    If so, what do you think the limiting value would be?

3. Let $f(x) = x \sin\left(\dfrac{1}{x}\right)$.

(a) Determine the range of fluctuation of f over each of the following deleted neighborhoods by 1) plotting the function and 2) analytically by evaluating the function at a finite number of random points as suggested in the previous problem.
    (i)   a symmetric interval about x=0 of length .2
    (ii) a symmetric interval about x=0 of length .02.
    (iii) a symmetric interval about x=0 of length .0002.

(b) Looking at the above results, would you say that the limit of f(x) at x=0 exists?
    If so, what do you think the limiting value would be?

LAB # 3 — *Continuity*

---

1. Consider each of the following functions defined below. For each of them, determine whether or not it is continuous at x=2. If not, determine the type of discontinuity there. If there is a removable discontinuity at x=2, how would you define the function at x=2 to make it continuous there?

(a) $f(x) = \dfrac{1}{x-2}$

(b) $g(x) = \begin{cases} x+3 & ,x<2 \\ x^2 - 2x + 5, & x>2 \end{cases}$

(c) $h(x) = \cos\left(\dfrac{1}{x-2}\right)$

(d) $k(x) = (x-2) \cos\left(\dfrac{1}{x-2}\right) + 3$

2. Let $f(x) = \begin{cases} -x^3 - x + 5 & ,x \leq 2 \\ cx^3 - 3x^2 + 9x - 25 & ,x > 2 \end{cases}$

(a) Is f(x) continuous at x=2 for c=1? If not, what kind of discontinuity is it?

(b) Is there a value of c such that f(x) is continuous on [-2,6]? Make sure that you can show on paper that your value of c is "as advertised".

(c) Is the resulting curve smooth? Why or why not? (Be sure to show the graph.)

(d) If there is a value of c as specified above, what kind of spline results? What is the set of data points?

(e) Show that f assumes the value π on the interval [0,2]. Then find a value of x in [0,2] such that $f(x)=\pi$.

(f) Show that f has a zero in [0,2] and then find one there.

3. A study has been done on heat loss from the body. This data suggests the following formula for the windchill index .

$$wci(v,F) = \begin{cases} F & ,0 \leq v < 4 \\ 91.4 - \dfrac{(10.45 + 6.69\sqrt{v} - 0.447v)(91.4 - F)}{22.042} & ,4 \leq v \leq 45 \\ 1.60F - 55 & ,v > 45 \end{cases}$$

where **v** is the wind speed in miles per hour, **F** is the air temperature in degrees Fahrenheit, and **wci** is the equivalent temperature in degrees Fahrenheit felt by exposed skin for the specified temperature and wind speed. This formula is based on a wind speed of 4 miles per hour being the standard condition. (William Bosch and L.G.Cobb,"Windchill",The UMAP Journal,Winter,1984, pp 480-492)

(a) For F constant, is wci(v,F) continuous? If not, where is it not continuous? (Use limits.)

(b) What is the windchill index for a temperature of 10 degrees Fahrenheit and a wind speed of 20 mi/hr?

(c) A city reports a windchill index of -90 degrees Fahrenheit and wind speed of 35 mi/hr. Find the air temperature.

(d) A city reports a windchill index of -65 degrees Fahrenheit and an air temperature of -10 degrees Fahrenheit. Find the wind speed.

(e) Plot the graph of wci(v,F) for F=20, 10 and 0. By examining the graphs, does wci(v,F) appear to be continuous for these values of F?
(NOTE. For F constant, wci is a function of v.)

LAB # 4 — *Secant Lines and Tangent Lines*

---

1. For each function, (i) compute an equation of the secant line passing through the indicated points; and, (ii) graph both the secant line and function on the indicated interval. The commands for the first problem are done for you.

(a) $f(x) = -1 - 3x - 5x^2 + 5x^3 + 5x^4$; Points: $(-1, f(-1))$ and $(1, f(1))$; Interval: $[-1.5, 1.5]$

Begin by defining f:
Clear[f];   **f[x_]=-1-3x-5x^2+5x^3+5x^4**

215

Then, f(-1) is **f[-1]** and f(1) is **f[1]**, so the slope of the secant line passing through the points (-1,f(-1)) and (1,f(1)) is obtained by

Clear[slope]

slope=(f[1]-f[-1])/(1-(-1))

Therefore, an equation of the line passing through the points (-1,f(-1)) and (1,f(1)) is obtained by

Clear[y]

y=slope (x-1)+f[1]

To graph f and y on the interval [-1.5,1.5], execute

Plot[{f[x],y},{x,-1.5,1.5},PlotStyle->{GrayLevel[0],GrayLevel[.3]}]

(b) $f(x) = -6 + x + 7x^3 + 5x^4 - 6x^5$; Points: $(-1, f(-1))$ and $(0, f(0))$; Interval: $[-1.5, 2]$

(c) $f(x) = \dfrac{-10 - 2x}{-9 + 4x}$; Points: $(2, f(2))$ and $(2.1, f(2.1))$; Interval: $[1, 3]$.

2. Let $f(x) = (-5 - 7x)\sqrt[3]{1 + 10x + 9x^2} = (-5 - 7x)(1 + 10x + 9x^2)^{1/3}$. Investigate

(a) $\displaystyle \lim_{h \to 0} \frac{f(-1 + h) - f(-1)}{h}$; and (b) $\displaystyle \lim_{h \to 0} \frac{f\left(\frac{-1}{9} + h\right) - f\left(\frac{-1}{9}\right)}{h}$.

(c) Graph f on the interval $[-1.25, 0]$.

3. Let $p(x) = \left| -160 - 188x + 120x^2 + 63x^3 \right|$.

(a) Graph p on the interval $[-3, 2]$

(b) Investigate $\displaystyle \lim_{h \to 0^+} \frac{p(a + h) - p(a)}{h}$, $\displaystyle \lim_{h \to 0^-} \frac{p(a + h) - p(a)}{h}$, and $\displaystyle \lim_{h \to 0} \frac{p(a + h) - p(a)}{h}$

for (i) $a = \dfrac{10}{7}$; (ii) $a = \dfrac{-2}{3}$; and (iii) $a = \dfrac{-8}{3}$.

LAB # 5 — *Slopes, Tangent Lines and Derivatives*

---

1. In each of the following exercises, use *Mathematica* to define the given function f, compute both **f'[x]** and **D[f[x],x]**, and graph f(x) and f'(x) on the indicated interval. In each case, express the answer as a fraction with a single denominator.

(a) $p(x) = 4x - 5x^2 - 5x^3 + x^4 + 2x^5 + 6x^6$; Interval: $[-1, 1]$.

(b) $f(x) = 2\cos(3x) - 4\sin(5x)$; Interval: $[0, 2\pi]$

2. Let u(x) and v(x) denote functions. If u'(x) and v'(x) exist for all values of x, compute the derivative of each of the following functions. The first problem is done for you.

(a) $f(x) = u(x)v(x)$

Clear[f]; f[x_]=u[x] v[x]

D[f[x],x]

(b) $f(x) = \dfrac{u(x)}{v(x)}$, $v(x) \neq 0$;

(c) $f(x) = u\big(v(x)\big)$;

(d) $f(x) = a\,u(x) + b\,v(x)$, a and b any two real numbers;

(e) $f(x) = u(x)^n$, n any non – zero number;

(f) $f(x) = \sin\big(u(x)\big)$; and

(g) $f(x) = \cos\big(v(x)\big)$.

3. For each function, (i) compute an equation of the line tangent to the graph of the function at the indicated point; and (ii) graph both the tangent line and the function on the indicated interval.

(a) $f(x) = \left(-9x + 2x^2\right)\sqrt[3]{1 - 3x - 5x^2}$ at $(1, f(1))$ on $[-2, 5]$.

(b) $p(x) = 2 + 7x^2 - 5x^3 - 6x^4 - x^5$ at $(-2, p(-2))$ on $[-4.75, 1.25]$.

(c) $t(x) = -2\cos(7x) - 7\sin(4x)$ at $\left(\dfrac{\pi}{3}, t\!\left(\dfrac{\pi}{3}\right)\right)$ on $[0, 2\pi]$.

4. If $p(x) = -1 - 7x^2 - 3x^3$, find open intervals on which (i) $p'(x)$ is positive; and (ii) $p'(x)$ is negative.

5. If $f(x) = 2\cos(3x) - \sin(7x)$, approximate open sub – intervals of $[0, \pi]$ on which (i) $f'(x)$ is positive; and (ii) $f'(x)$ is negative.

6. Find numbers a and b so that $g(x) = \begin{cases} 5 - x - 2x^2 + 5x^3, & \text{if } x > 0 \\ a + bx, & \text{if } x \leq 0 \end{cases}$ is differentiable when $x = 0$.

LAB # 6 — *Composition of Functions and The Chain Rule*

1. A runner on a straight track is being tracked by an automatic camera located 15 feet from the track. The camera is set up so that the operator can obtain information about the angle theta that the camera must be turned in order to keep trained on the runner as well as how fast the camera is turning in order to keep the runner in view. If, at a given instant, the camera is turning at a constant rate of 1.5 degrees per second, how fast is the runner going if the angular reading for theta is 32 degrees?

2. A variant on the above problem is to allow the track to be circular. In this case, the question of how fast the runner is going would translate into the question of a rate of change of arc length with respect to time. Thus we have the following setup.

Given: dthetat = d(theta)/dt = 9.8 degrees per minute

f = (1,0) - a base point on the circle

c = (1.5,1.5) - location of the camera

t0 = .4 minutes - length of time the camera has been in operation.

thetato = 39.2 degrees radians - the angle of the camera from the base position at time t=t0.

Find: d(arc length)/dt - the speed of the runner on the circle.

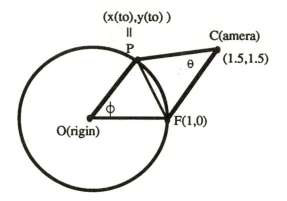

Observation:

Since the arc length s = phi*radius, then ds/dt = r*d(phi)/dt. Thus, we need to know d(phi)/dt. But phi = ArcTan[y/x] and y is a function of x, so d(phi)/dt = f(x'[t] , x(t)) (i.e. d(phi)/dt can be written as a function of x(t) and x'(t)). Thus all we need to do is to find x(t) at t0 and x'(t) at t0.

An outline of a mode of attack for this problem is:

1. Express the coordinates of the point P as a function of x. (Use the fact that P is on the circle.)

2. Express x(t) as a function of the the angle θ. Use the Law of Cosines applied to the triangle FPC.

3. Differentiate the expression obtained from the Law of Cosines. Solve for x(t) at t=t0.

LAB # 7 — *Related Rates and Implicit Differentiation*

---

1. Van der Waal's equation of state for a gas is

$$\left(P + \frac{a}{V^2}\right)(V - b) = R\,T$$

written for one mole of gas, where a, b and R are constants depending on the gas and P, T and V represent pressure, temperature, and volume respectively. Suppose the pressure is held constant while the temperature and volume are allowed to vary. Find the rate of change of the volume with respect to the temperature. (HINT. Use implicit differentiation.)

2. An object that weighs w lbs. on the surface of the earth weighs approximately

$$\text{weight}(r) = w \left(\frac{3960}{3960 + r} \right)^2 \text{ lb}$$

when lifted a distance of r miles from the earth's surface. Find the rate at which the weight of an object, which weighs w lbs. on the earth's surface, is changing when it is r miles above the earth's surface and is being lifted at the rate of v mi/sec, for the given values of w, r and v.

(a) w = 100, r = 75, v = 10   (b) w = 2000, r = 125, v = 15

3. Oil is leaking from an ocean tanker at the rate of 4000 liters/sec. The leakage results in a circular oil slick with a depth of 6 cm.

(a) How fast is the radius of the oil slick increasing when the radius is 50 miles?

(b) How fast is the radius of the oil slick increasing 5 hours after the spill began?

4. Boyle's law for confined gases states that if temperature is constant, then pv = constant, where p is the pressure and v is the volume. At a certain instant, the volume is 95 in^3, the pressure is 25 lb / in^2, and the pressure is increasing at the rate of 1.5 lb / in^2 per minute. At that instant, what change is being effected on the volume?

5. When two resistors R_1 and R_2 are connected in parallel, the total resistance R is given by the equation $\frac{1}{R} = \frac{1}{R_1} + \frac{1}{R_2}$. If R_1 and R_2 are increasing at rates of .015 ohms / sec and .0175 ohms / sec, respectively, at what rate is R changing at the instant that $R_1 = 25$ ohms and $R_2 = 72$ ohms?

6. The graph of $x^3 + y^3 = 15xy$ is called a folium of Descartes.

(a) Plot the graph.
(b) Find yprime and slope as in EXAMPLE 7.
(c) Find the equation(s) of the tangent line(s) to the graph corresponding to x = 3.

LAB # 8 — *Graphing Functions Using Properties of the Derivative*

1. Discuss the behavior of the given function on the indicated interval.

(a) $g(x) = 4x^6 - 6x^4$, Interval $[-3, 3]$;

(b) $f(x) = \frac{2x - 3}{x^2 - 1}$, Interval $[-3, 1]$;

(c) $h(x) = \frac{21 - 7x + 2x^2 + 3x^3}{21 - 45x^3 + 4x^5}$, Interval $(-\infty, \infty)$;

(*Mathematica* denotes the symbol ∞ by **Infinity**);

(d) $g'(x) = -1.08813 \cdot 10^9 + 9.72273 \cdot 10^8 x - 2.11294 \cdot 10^8 x^2$
$$- 1401930 x^3 - 1568 x^4 + 2x^5,$$

Interval $(-\infty, \infty)$

(*Mathematica* denotes the symbol ∞ by **Infinity**);

(e) $k(x) = \dfrac{\sqrt[5]{x^3(3-7x)^4}}{5}$, Interval $[-1,1]$.

2. Locate the subinterval(s) of $[0,\pi]$ for which $f(x)=\sin(3x)+2\cos(2x)$ is (i) increasing and concave up; (ii) increasing and concave down; (iii) decreasing and concave up; and (iv) decreasing and concave down. In addition locate and classify all critical points. Neatly sketch, label and interpret each result on your paper.

3. (a) Find conditions for a, b, c, and d so that $f(x)=\dfrac{a}{3}x^3 + \dfrac{b}{2}x^2 + cx + d$

     is (i) always increasing; and (ii) always decreasing.

(b) Use (a) to choose a, b, c, and d so that $f(x)=\dfrac{a}{3}x^3 + \dfrac{b}{2}x^2 + cx + d$ is

     always increasing. Verify your choices by graphing the result.

(c) Use (a) to choose a, b, c, and d so that $f(x)=\dfrac{a}{3}x^3 + \dfrac{b}{2}x^2 + cx + d$ is

     always decreasing. Verify your choices by graphing the result.

LAB # 9 — *Applied Max/Min Problems (one variable)*

---

1. A bee's cell can be modelled as a regular hexagonal prism with one open end and one trihedral apex. Let's be more specific, and less erudite. We may construct this surface as follows.
   (a) Start with a regular hexagonal base of side s as in fig. 1 below.
   (b) Over this base, raise a right prism of a certain height h and with top ABCDEF as in fig. 3 below.
   (c) Cut off the corners A, C, and E by planes through the lines BD, DF, and FB. (See fig. 3.) The three planes all intersect at a point V.
   (d) The three cut-off pieces are tetrahedrons. See fig. 3 where the corner cut off at E is shown.
   (e) Put these three pieces on top of the remaining solid such that X, Y, and Z coincide with V. (See fig. 4.) Thus the lines BD, DF, and FB act as hinges around which the corners are rotated.

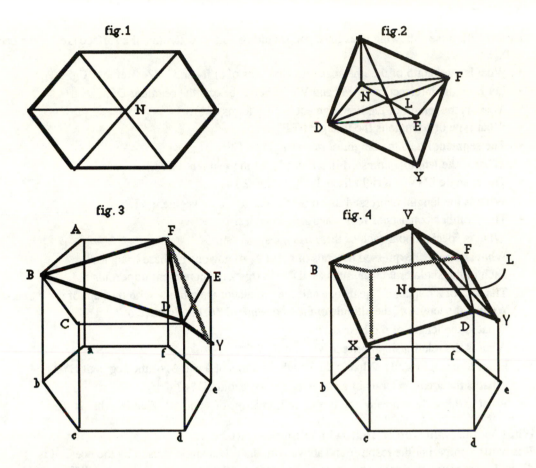

fig.1

fig.2

fig. 3

fig. 4

Some observations.

(a) The faces BXDV, VDYF, and BVFZ are rhombuses (parallelograms with any two adjacent sides equal).

(b) The volume of the bee's cell, the final surface shown in fig. 4, is the same as the volume of the original surface, the surface shown in fig. 3.

(c) The line segment VN goes through the center of the base abcdef and is perpendicular to the base.

(d) The diagonals of a rhombus are perpendicular to each other.

The bee's form the faces by using wax. When the volume is given, it is economical to spare wax, and therefore, to choose the angle of inclination of the planes that cut off the corners, the angle

NVY, in such a way that the surface of the bee's wax is minimized. Thus, the problem we want to consider is:

Given: s = length of any side of the base.

h = the height of the right prism used to construct the bee's cell = length of aD.

Find: the angle of inclination θ = angle NVY that minimizes the surface area.

There are two things that we need to do. **First**, we need to find a function of θ, f(θ), that describes the surface area of the bee's cell. **Second**ly, we need to find the minium value of f(θ).

O   **Finding f(θ).**

By answering the following questions, you should be able to determine the surface area of the bee's cell.
- What is the length of the line segments (in terms of s) fN and Nd? (Refer to fig. 1.)
- Let L be the intersection of FD and VY. Then L bisects the segment DF.
  What is the length of the line segment NL (in terms of s)?
- What type of triangle is the triangle NFE?
- The segment FL is the height of the triangle NFE.
  What is the length, expressed in terms of s, of the segment FL?
- The triangle LNV is a right triangle. (See fig. 2.)
  What is the length, expressed in terms of s and θ, of the segment VL?
- The rhombus consists of four congruent triangles.
  What is special about each of these triangles?
  What is the area, expressed in terms of s and θ, of any one of these triangles?
- Each of the six lateral faces, such as dDYe in fig. 4, are congruent trapezoids.
  The area of a trapezoid equals one half the product of the altitude with the sum of the bases.
- What is the length of the altitude of the trapezoid dDYe, the segment de?
- What is the length of the base dD?
- What is the relationship between VN and EY?
- Refer to triangle VNL. What is the length, in terms of s and θ, of the segment EY?
- What is the area , in terms of s and θ, of the trapezoid dDYe?
- What is the surface area of the bee's s cell, in terms of s and θ? That is, what is f(θ) ?

O **What is the value of θ that will minimize   f(θ)?**
It is worth comparing the result found above with the actual angle chosen by the bees. It is difficult to measure this angle, however, the average of all measurements does not differ significantly from the theoretical value found above.

LAB # 10 — *Antidifferentiation*

---

1. Suppose that u(x) and v(x) are antiderivatives of w(x).
    (a) How are u(x) and v(x) related algebraically?
    (b) How are u(x) and v(x) related graphically?
    (c) What is u'(x)? v'(x)?

2. Let $f(x) = x^2 \cos(x)$.

    (a) Find $\int f(x)dx$.

    (b) Check your answer by an appropriate differentiation.

    (c) Plot three antiderivatives of f(x) simultaneously on $[0, 4\pi]$.

(d) How do these three antiderivatives relate to each other algebraically? Graphically? Be specific!

3. Let $g(x) = \dfrac{x^2 + 1}{(x + 2)(x^2 - 4)(x^2 - 1)}$.

   (a) Find $\int g(x)dx$.

   (b) Check your answer with an appropriate differentiation.

   NOTE. You may need to use the **Together** command with your verification.

   (c) Plot $g(x)$ on $[-3,3]$.

4. When discussing the motion of a point on a straight line, the functions of time s(t) (position), v(t) (velocity), and a(t) (acceleration) are related by the equations a(t)=v'(t) and v(t)=s'(t). Find v(t) and s(t) if a(t)=-3sin(t), v(π)=-2, and s(π)=-1.

5. What constant acceleration will enable the driver of a car to increase his speed from 10 mph to 70 mph in 8 seconds?

6. Solve the Boundary Value Problem y"=2sin(2x) - 4x+3, y'(π)=2, and y(π)=5.

7. At each point (x,y) on a curve, y"=3x-1 and the tangent line to the curve at the point (1,-3) is y=2-5x. Find the equation of the curve if the following point is on the curve:
   (a) (-1,4)
   (b) (2,π)

8. Suppose a beam of length L is at rest with a support under each end. If there is a uniform load on the beam, its shape is determined from the differential equation $y'' = kLx - kx^2$, where k is a constant. Find y if $y'\left(\dfrac{L}{2}\right) = 0$ and $y(L) = 0$.

LAB # 11 —Riemann Sums

---

1. Let f(x)=x+sin(πx). It is a fact that

   $\int_1^3 f(x)\,dx = 4$. Find a positive integer N such that whenever n is any positive integer greater than or equal to $N, |4 - \mathbf{riemannsum[f[x], \{x, 1, 3\}, n]}| < .001$.

2. The area of a circle with radius r is $\text{Area}_{circle} = \pi r^2$. Therefore, the area of one – fourth of a circle is $\dfrac{\pi r^2}{4}$. Notice that the equation of the unit circle is given by $x^2 + y^2 = 1$. The function $y(x) = \sqrt{1 - x^2}$ corresponds to the upper half of the unit circle. Hence, $\int_0^1 \sqrt{1 - x^2}\,dx = \dfrac{\pi}{4}$.

Find a positive integer N such that whenever n is any positive integer greater than or equal to N,

$$\left| \frac{\pi}{4} - \text{riemannsum}[y[x], \{x, 0, 1\}, n] \right| < 10^{-10} .$$

3. (Sums of Integrals) Suppose f and g are integrable functions on the interval [a,b]. Use Riemann sums to **compare** the two integrals

   (i) $\displaystyle\int_a^b (f(x) + g(x)) \, dx$ and (ii) $\displaystyle\int_a^b f(x) \, dx + \int_a^b g(x) \, dx$.

4. (Betweenness) Suppose f is an integrable function on the interval [a,b]. Let c be a point between a and b. Use Riemann sums to compare the two integrals

   (i) $\displaystyle\int_a^b f(x) \, dx$ and (ii) $\displaystyle\int_a^c f(x) \, dx + \int_c^b f(x) \, dx$.

5. Suppose f and g are integrable on [a,b] and f(x) > g(x) on [a,b]. Use Riemann sums to compare the two integrals

   (i) $\displaystyle\int_a^b f(x) \, dx$ and (ii) $\displaystyle\int_a^b g(x) \, dx$.

LAB # 12 —*Area*

---

1. Find the area of the region(s) bounded by the curves

$$y = 1 - \frac{x^2}{2} + \frac{x^4}{4} \ , y = \cos(x) \, , x = \frac{-\pi}{2}, \text{ and } x = \frac{\pi}{2} .$$

2. Let k be a number bigger than 0 and consider the region bounded by the graph of

$$x^2 + \frac{y^2}{k} = 2 - k.$$

What value of k maximizes the area of this region?

HINTS: 1. Solve for y in terms x (and k). Taking the non-negative value for y, you get a function f(x,k).
  2. Plot f(x,k) and -f(x,k) for two or three different values of k and then graph these curves simultaneously using the **Show[*,*,*]** command. What's the significance of the picture obtained from this process?
  3. Define a function area(k), in terms of an integral, that gives the area of the bounded region.
  4. Plot area(k) on an appropriate interval to get a "feel" for the function.
  5. Using Differential Calculus, find where the maximum area is achieved and the value of the maximum area.

3. Let $f(x) = e^{-(x-3)^2 \cos\left(\frac{2p x}{3}\right)} - 10$ and $g(x) = 3 \cos(x - 3)$ on the interval $[1,5]$. Use the command FindRoot to locate the intersection points. Then approximate the total area of the region (s) bounded between the graphs of f and g.

4. A typical problem in thermodynamics is to find the work done by an ideal Carnot engine. This work is defined as the area of the region defined by the isothermals $xy = a$ and $xy = b$, where $0 < a < b$, and the adiabatics $xy^{1.4} = c$ and $xy^{1.4} = d$, where $0 < c < d$. Find the work done by a Carnot engine by finding the area of the region R in the following diagram.

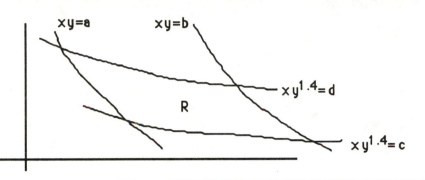

LAB # 13 —*The Fundamental Theorem of Calculus*

---

1. Let $h(x) = \cos(x)$ on $[0, 2\pi]$ and let $caph(x) = \int_0^x h(t)\,dt = \int_0^x \cos(t)\,dt$. Graph caph on the interval $[0, 2\pi]$. Visually, determine a closed form for caph. What is the derivative of caph?

2. Define and graph $f(x) = x^2 + \frac{1}{x^2}$ on the interval $[1,4]$. Use the command **Integrate** to compute an antiderivative of $f(x)$.

   (a) Does the Fundamental Theorem of Calculus apply to f on the interval $[1,4]$?

   (b) Use the **Integrate** command to compute the value of $\int_1^4 f(x)\,dx$.

3. Define and graph $g(x) = \sin^2(x) \cos^3(x)$ on the interval $\left[0, \frac{\pi}{2}\right]$. Use the command **Integrate** to compute an antiderivative of $g(x)$.

   (a) Use the **Integrate** command to compute the value of $\int_0^{\pi/2} g(x)\,dx$.

   (b) Use the **Integrate** command to compute the value of $\int_1^4 g(x)\,dx$.

225

4. Define and graph $h(x) = \dfrac{x}{(x-1)^3}$ on the interval $[0,2]$. Use the command **Integrate** to compute an antiderivative of $h(x)$.

(a) Use the **Integrate** command to compute the value of $\int_0^2 h(x)\,dx$.

(b) Does the Fundamental Theorem of Calculus apply to h on the interval $[0,2]$?

5. (a) Carefully enter the following commands. Explain each result.

   Plot[Sin[x]/x,{x,0,8Pi},PlotRange->All]

   Together[D[Sin[x]/x,x]]

(b) Type and enter the command

   Limit[Sin[x]/x,x->0]

and use the result to compute $\lim\limits_{x \to 0} \dfrac{\sin(x)}{x}$.

(c) Evaluate $\displaystyle\int_0^{8\pi} \dfrac{x\cos(x) - \sin(x)}{x^2}\,dx$.

LAB # 14 — *The Mean Value Theorem For Integrals*

---

1. Find numbers that satisfy the conclusion of the Mean Value Theorem for Integrals for each of the following integrals:

(a) $\displaystyle\int_0^1 \dfrac{1}{\sqrt{1+x^2}}\,dx$

(b) $\displaystyle\int_2^3 \sqrt{1+x^3}\,dx$

(c) $\displaystyle\int_0^\pi \sin\sqrt{x}\,dx$

2. Let T denote triangle ABC with AB=8 and BC=11. What must the length of CA be in order to maximize the area of T? In order to answer this question, do each of the following:
(a) Plot area(x) on $[3,19]$. (Refer to EXAMPLE 6.)
(b) Notice that graphically, area(x) has a maximum on the interval $[3,19]$. Then, find the value of x in the interval where area(x) has maximum value.
   HINT. Find where area'(x)=0.

3. Let T denote triangle ABC with AB=1 and BC=4.
(a) What must the length of CA be in order to yield the average value of the area of T?
(b) What must the length of CA be in order to maximize the area of T?

4. Let s(t) be the position function of a particle moving on a coordinate line. For velocity function v(t) and acceleration function a(t), it has been shown earlier that s'(t)=v(t) and v'(t)=a(t). The

average velocity v_{ave} over $[t_0,t_1]$ is defined to be $v_{ave} = \dfrac{\text{change in position}}{\text{change in time}} = \dfrac{s(t_1) - s(t_0)}{t_1 - t_0}$.

Now, $\displaystyle\int_{t_0}^{t_1} v(t)dt = s(t)\Big|_{t_0}^{t_1} = s(t_1) - s(t_0)$ and hence $\dfrac{1}{t_1 - t_0} \displaystyle\int_{t_0}^{t_1} v(t)dt = v_{ave}$. That is, v_{ave} is

the average value of v(t) over $[t_0,t_1]$.

Similarly, it can be shown that a_{ave} is the average value of a(t) over $[t_0,t_1]$.

Given that v(t)= 16t+1 on [0,6], find:

(a) v_{ave} and

(b) a_{ave}.

LAB # 15 —*Volumes of Solids of Revolution*

---

1. Let $f(x) = x \sin(x)$ and let R denote the region bounded by the graphs of the x − axis, f, x = 0,

and $x = \dfrac{\pi}{2}$. Find the volume of the solid obtained by

(a) revolving R around the x-axis. Visualize this solid using **solidrev.**
(b) revolving R around the y-axis. Visualize this solid using **solidrev.**

2. Let $f(x) = \dfrac{x^5}{60} - \dfrac{x^3}{3} + 2x + 1$ and $g(x) = \dfrac{7x}{20} + 1$. Let R denote the region in the first quadrant

bounded by f and g, x = 1, and x = 4. Find the volume of the solid generated by

(a) revolving R around the x-axis. Visualize this solid using **solidrev.**
HINT. Let R(1) denote the region bounded by the x-axis, the graph of f, x=1, and x=4. Let R(2) denote the region bounded by the x-axis, the graph of g, x=0 and x=3. First find the volume of the solid generated by revolving R(1) around the x-axis (call it V(1)); then find the volume of the solid generated by revolving R(2) around the x-axis (call it V(2)). Then the desired volume is V(1)-V(2).

(b) revolving R around the y-axis. Visualize this solid using **solidrev.**

3. Let $f(x) = 4x^6 - 20x^5 + 37x^4 - 30x^3 + 9x^2$ and let R denote the region bounded by the x − axis,

the graph of f, x = 0, and $x = \dfrac{3}{2}$. Find the volume of the solid generated by:

(a) revolving R about the x-axis;   (d) revolving R about the line x=-k, k>0;
(b) revolving R about the y-axis;   (e) revolving R about the line y=-1; and
(c) revolving R about the line x=-2;   (f) revolving R about the line y=-k, k>0.

4. Let $f(x) = \cos(x - 2) + \dfrac{1}{2}$ and $g(x) = e^{-(x-2)^2} \cos(\pi(x - 2)) + 1$ on [0,4].

Let R denote the region bounded by the graphs of f and g.

(a) NUMERICALLY compute the volume of the solid generated by revolving R about the x – axis;

(b) NUMERICALLY compute the volume of the solid generated by revolving R about the y – axis.

5. Let $0 < k < 1$ and let $g_k(x) = \dfrac{\sin(kx)}{1-k}$.

(a) Verify that $g_k(x)$ is positive and continuous on the interval $\left[0, \dfrac{\pi}{k}\right]$.

(b) Let V_k denote the volume of the solid obtained by revolving the region bounded by the graphs

of $g_k(x)$, $x = 0$, $x = \dfrac{\pi}{k}$, and the x – axis around the x – axis. What value of k minimizes V_k? Sketch the resulting solid.

6. Let f and g be continuous non-negative functions on [a,b]. Let R(f) denote the region bounded by the x-axis, the graph of f, x=a, and x=b. Similarly, let R(g) denote the region bounded by the x-axis, the graph of f, x=a, and x=b. Let R be the region bounded by f, g, x=a, and x=b. Let **V(f)** be the volume of the solid generated by revolving R(f) around the x-axis; **V(g)** be the volume of the solid generated by revolving R(g) around the x-axis; and **V** the volume of the solid generated by revolving R around the x-axis.

State conditions such that the following hold. Illustrate these conditions by a sketch.
(a) V=V(f)-V(g)
(b) V=V(g)-V(f)
(c) V is neither of these.

LAB # 16 — *Work = Force through Distance*

---

1. A cylindrical tank of radius 3 feet and length 11 feet is lying on its side on horizontal ground. If this tank initially is full of gasoline weighing 40 pounds per cubic foot, how much work is done in pumping this gasoline to a point five feet above the top of the tank?

Questions you may want to ask:

Partition the interval $[0,11]$ into a finite number of points h_i and imagine slicing the liquid volume by horizontal plates into slabs of thickness Δh_i.

What is the volume of the i - th slice?

How far will the i - th slice have to be lifted?

What is the work necessary to lift the i - th slice?

2. Coulomb's Law governing the force of attraction between two electrically charged bodies states that the force of attraction or repulsion between two point charges is directly proportional to the product of the charges and inversely proportional to the square of the distance between them, that is,

$F = k \dfrac{q_1 \, q_2}{d^2}$, where q_1 and q_2 are the magnitudes of the charges at points P_1 and P_2, respectively, d is the distance between P_1 and P_2, and k is the constant of proportionality. Suppose that charges of equal magnitude are 5 meters apart and that the force with which the two charges repel each other is 10 newtons.

(a) What is the value of k in terms of q ?

(b) How much work is required to bring one charge from P_2 to a distance of 2 m from P_1 ?

(c) Suppose that an ion with a charge of +1 is stationary and located at (5,0), and a second ion with charge +1 is located at (-8,0) and is stationary. How much work is required to move a third ion with charge +1 from (0,0) to (3,0)?

3. A right circular conical tank of altitude 20 feet and base radius of 5 feet has its vertex at ground level and axis vertical. If the tank is full of water, find the work done in pumping the water over the top of the tank.

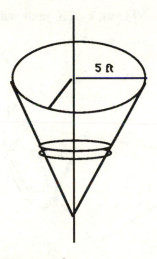

Questions you may want to ask:
What is the volume of the i-th slice?
How far will the i-th slice have to be moved?

4. Suppose it took twenty years to construct the great pyramid of Knufu at Gizeh. This pyramid is 500 feet high and has a square base with edge length 750 feet. Suppose also that the pyramid is made of rock with density 200 pounds per cubic foot. Likewise, suppose each laborer did 160 foot pounds per hour effective work in lifting rocks from ground level to position in the pyramid and worked 12 hours daily for 330 days per year. How many laborers were required to construct this pyramid?

LAB # 17 — *Inverse Functions*

---

1. Find the value of x that maximizes the angle a in the figure at the top of the next page. How large is a at that point?

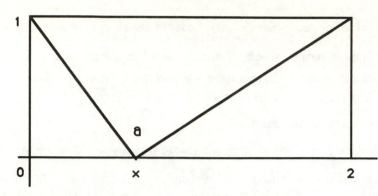

2. As shown in the figure, a sailboat is following a straightline course L. The shortest distance from a tracking station T to the course is D miles. As the boat sails, the tracking station records its distance k from T and its direction θ with respect to T. Angle α specifies the direction of the sailboat.

   (a) Express α in terms of d, k, and θ.

   (b) Estimate α to the nearest degree if d=50 yards, k = 210 yards and θ = 53.4 degrees.

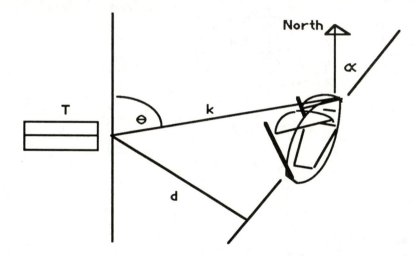

3. Let $f(x) = \int_0^x \sin^4(t^2)dt$. Show that $f(x)$ has an inverse.

4. Let g(x)=(1-x) arcsin(x) on the interval [0,1]. Find the maximum value of g(x) on the interval.

5. Let R denote the region bounded by the graphs of $y = \dfrac{1}{\sqrt{1+x^2}}$, x = 0, x = 1, and the x - axis.

   (a) Use cylindrical shells to find the volume of the solid obtained by revolving R around the y-axis.
   (b) Use the disk method to find the volume of the solid obtained by revolving R around the x-axis.

LAB # 18 — *Logarithmic / Exponential Functions & Rates of Growth*

---

1. Plot the graphs of $f(x) = \ln(x)$, $g(x) = e^x$, and $h(x) = x$ on the same axes. Distinguish between the graphs.

2. Plot the graphs of $f(x) = e^x$, $g(x) = e^{-x}$, and $h(x) = e^x \cos(x)$ on the same axes. Distinguish between the graphs.

3. Let $q(x)$ denote any polynomial. Using similar procedures as in the example, calculate each of the following limits.

   (a) $\displaystyle \lim_{x \to \infty} \frac{q(x)}{\ln(x)}$;

   (b) $\displaystyle \lim_{x \to \infty} \frac{e^x}{q(x)}$; and

   (c) $\displaystyle \lim_{x \to \infty} \frac{q(x)}{e^x}$.

4. Sketch the graph of each of the following functions on the indicated interval. Find and classify all relative extrema and points of inflection.

   (a) $f(x) = 1 - e^{-x^.}$, Interval $(-\infty, \infty)$;

   (b) $g(x) = e^x \cos(x)$, Interval $[-2p, 4p]$.

5. Let R denote the region bounded by the graphs of $y = e^{-x^2}$, $y = 0$, $x = 0$, and $x = 1$. Find the volume of the solid generated by revolving R about the $y-$axis.

6. The region bounded by the graphs of $y = \dfrac{1}{\sqrt{x}}$, $y = 0$, $x = 4$, and $x = 9$ is revolved about the $x-$axis. Find the volume of the resulting solid.

LAB # 19 — *Improper Integrals*

---

1. If interest is compounded m times per year, then the present value of profits for a business over n years is $P(n,m) = \displaystyle\int_0^n \left(1 - \frac{r}{m}\right)^{mt} pr(t)\,dt$, where $pr(t)$ is the expected profit per year and r is the interest rate.

   Assuming $r = .85\%$ and expected profits are \$225,000 per year, find the present value of profits for a business over a 20 year period if interest is compounded:
   (a) yearly   (c) quarterly
   (b) semianually   (d) monthly

❑ EXERCISES ❑

If the interest is compounded continuously, then the formula for the present value after n years

becomes $C(n) = \int_0^n e^{-rt} pr(t)\, dt$.

(e) Calculate C(20) and compare C(20) with the results above.

(f) Find $\lim_{n \to \infty} C(n)$, $\lim_{n \to \infty} P(n,1)$, $\lim_{n \to \infty} P(n,2)$, $\lim_{n \to \infty} P(n,4)$, and $\lim_{n \to \infty} P(n,12)$.

These values can be interpreted as the present value of all future anticipated profits for each respective interest rate.

(g) What type of interest would you want to receive?

2. Using the formula introduced in the above problem (i.e. the present value of future profits is given

by the integral $\int_0^\infty e^{-rt} pr(t)\, dt$, where r is the interest rate and pr(t) is the expected profit t years

from now), explain why the present value of a company's future profits goes down as interest rates go up. (Thus, investors tend to find savings accounts more attractive than future profits of a company when rates go up).

3. The reliability R(t) of a product is the probability that it will not require repair for at least t years. To design a warranty guarantee, a manufacturer must know the average time of service before repair

of a product. This is given by the improper integral $\int_0^\infty (-t) R'(t)\, dt$ for many high – quality

products where R(t) has the form e^{-kt} for some positive constant k . Assuming this form for R(t), find an expression (in terms of k) for the average time of service before repair.

4. A space capsule weighing 1000 pounds on the surface of the earth is propelled out of earth's gravitational field. How much work is done against the force of gravity? By Newton's

inverse - square law, the force necessary to offset gravity is $F(x) = \dfrac{k}{x^2}$ where x represents

the distance of the capsule from the center of the earth. Using x = 4000 as the radius yields

$k = 16 \times 10^9$. The work done in sending the capsule out of earth's gravitational field is given

by the integral $\int_0^\infty F(x)\, dx$. What is the value of this integral?

5. Suppose the universe had been made differently and the force of gravity was $F(x) = \dfrac{k}{x}$

instead of $F(x) = \dfrac{k}{x^2}$. What ramifications would this have for our being able to put a

capsule into space?

6. Consider the surface generated by revolving the curve y = 1/x, 1≤ x <∞ , about the x-axis. This surface is called **Gabriel's horn**.
   (a) Plot the curve's y = 1/x and y=-1/x over the interval [1,50]. Then change the interval to [1,100] and replot the graph. Use the option **PlotRange -> All** to see the entire graph.

   (b) Find the area between the two curves, 1≤ x <∞.

232

(c) Find the volume enclosed by Gabriel's horn.

(d) Find the surface area of Gabriel's horn.

(e) Logically, do the answers make sense to you? Why or why not? Explain.

LAB # 20 — *Monotone Sequences and Rates of Growth*

---

1. Let $a_n = \ln(100\,n)$ and $b_n = n^3$.

   (a) Define the sequences.

   (b) Create a table **tbl** of values of $\dfrac{a_n}{b_n}$ for $n = 1$ to 100.

   (c) Plot the table of ratios with respect to n.

   (d) Based on this information, compare rates of growth and make an educated guess about

$$\lim_{n \to \infty} \frac{a_n}{b_n} \; .$$

   (e) How large would n have to be in order that $\dfrac{a_n}{b_n} < .001$?

   (f) After taking an appropriate limit, compare the rate of growth of a_n with that of b_n.

2. Let $a_n = n^{100}$ and $b_n = 2^n$.

   (a) How large would n have to be before $\dfrac{a_n}{b_n} < 1$? .01? .00001?

   (b) Notice that initially, a_n is larger than b_n. Is $\left\{ \dfrac{a_n}{b_n} \right\}$ a monotone sequence? If yes, is it

   monotone increasing or monotone decreasing? Justify your answer.

   (c) How far out do you have to go before $\left\{ \dfrac{a_n}{b_n} \right\}$ becomes decreasing? Give an argument to show

   that once $\left\{ \dfrac{a_n}{b_n} \right\}$ becomes decreasing, then it will continue to decrease from that point on.

   (d) Compare the rates of growth of the two sequences.

3. Discuss how the answers to the above questions would change if we let $b_n = 9^n$.

4. Let $a_n = 17^n$ and $b_n = n!$ How large would n have to be in order that $\dfrac{a_n}{b_n}$ be

   less than 1? .001? .000001?

5. Let $a_n = n^n$ and $b_n = 5^{n^2}$. How large would n have to be in order that $\dfrac{a_n}{b_n}$ be

   less than 1? .001? .000001?

LAB # 21 — *Series*

---

1. Evaluate each series:

(a) $\displaystyle\sum_{n=0}^{\infty} \frac{12^{n+2}}{13^{n-1}}$

(b) $\displaystyle\sum_{n=0}^{\infty} \frac{2}{4n^2 - 1}$

2. Find all values of x for which each series converges.

(a) $\displaystyle\sum_{n=0}^{\infty} \frac{1}{\sin^n(x)}$

(b) $\displaystyle\sum_{n=0}^{\infty} \frac{1}{(x-1)^n}$

(c) $\displaystyle\sum_{n=0}^{\infty} \frac{1}{\left(x^2 - 3x - 3\right)^n}$

(d) $\displaystyle\sum_{n=0}^{\infty} \frac{1}{\cos^2(x) - \sin^2(x)}$

3. Find an upper bound for the area bounded by the graph of $f(x) = \cos(x^2)$ and the x - axis for $0 \le x \le \infty$.

LAB # 22 — *Taylor Polynomials*

---

1. Let f(x) = sin(x) and a = 0.

  (a) Calculate taylorpoly3f(x), taylorpoly4f(x), taylorpoly10f(x), and taylorpoly20f(x).
  (b) Graph the triple {f(x), taylorpoly3f(x), taylorpoly4f(x)} on one graph and the triple {f(x), taylorpoly10f(x), taylorpoly20f(x)} on a second graph over the interval [-π,π].
  (c) Calculate errorn(x) = f(x) - taylorpolynf(x) for n=3,4,10, and 20.
  (d) Graph the pair {error3(x), error4(x)} over the interval [-π,π]. Graph the pair {error10(x), error20(x)} over the interval [-π,π].
  (e) What is the maximum error made in approximating f(x) by taylorpolynf(x), n=3,4,10, and 20, over the interval [-π,π]?
  (f) What is the smallest value of n for which taylorpolynf(x) would approximate f(x) to within a 3 place accuracy, |f(x) - taylorpolynf(x)| ≤ .0005, over the interval [-π,π]?

2. Let $f(x) = \text{Tan}^{-1}(x)$ and $a = .5$.

(a) Calculate taylorpolynf(x) for n = 3, 4, 10, and 20.

(b) Graph the triple {f(x), taylorpoly3f(x), taylorpoly4f(x) } on one graph and the triple {f(x), taylorpoly10f(x), taylorpoly20f(x)} on a second graph.

(c) Calculate errorn(x) = f(x) - taylorpolynf(x) for n=3,4,10, and 20.

(d) Graph the pair {error3(x), error4(x)} on the interval [-1.5,2.5]. Graph the pair {error10(x), error20(x)} on the interval [-1.5,2.5]. Change the plot interval and regraph the functions so that the maximum absolute error on the interval is no larger than .01.

(e) Complete the entries in the following table.

n	error	expansion point	maximum interval for which $|g(x) - \text{taylorpolyng}(x)| \leq$ error	interval width
2	.0005	.5		
3	.0005	.5		
10	.0005	.5		
20	.0005	.5		

3. Let r(x) be the rational function $r(x) = \dfrac{x^3 + 9x^2 + 36x + 60}{3x^2 - 24x + 60}$.

(a) Graph $f(x) = e^x$ and r(x) over the interval $[-1,1]$.

(b) Let error(x) = f(x) - r(x). Graph the function error(x) over the interval [-1,1].

(c) Find the maximum value of the function error(x) over the interval [-1,1].

(d) What Maclaurin Polynomial approximates f(x) with the same accuracy as r(x) over the interval [-1,1]?

LAB # 23 — *Approximations by Taylor Polynomials*

---

1. Let $f(x) = \sin(x)$ and consider the Taylor Series expansions about $x = \dfrac{\pi}{2}$.

(a) Find the Taylor Polynomials taylorpoly1f(x), taylorpoly3f(x), taylorpoly5f(x), and taylorpoly7f(x) .

(b) Plot f(x), taylorpoly1f(x), taylorpoly3f(x), taylorpoly5f(x), and taylorpoly7f(x) on $\left[\dfrac{\pi}{2} - 2\pi, \dfrac{\pi}{2} + 2\pi \right]$.

(c) Plot f(x), taylorpoly1f(x), taylorpoly3f(x), taylorpoly5f(x), and taylorpoly7f(x) on $\left[\dfrac{\pi}{2} - \pi, \dfrac{\pi}{2} + \pi \right]$.

(d) Plot $f(x)$, taylorpoly1f(x), taylorpoly3f(x), taylorpoly5f(x), and taylorpoly7f(x) on

$$\left[\frac{\pi}{2} - \frac{\pi}{4}, \frac{\pi}{2} + \frac{\pi}{4} \right].$$

(e) Plot $f(x)$, taylorpoly1f(x), taylorpoly3f(x), taylorpoly5f(x), and taylorpoly7f(x) on

$$\left[\frac{\pi}{2} - \frac{\pi}{8}, \frac{\pi}{2} + \frac{\pi}{8} \right].$$

(f) What conclusions can you draw from the above?

(g) Find the Taylor Polynomials taylorpoly2f(x), taylorpoly4f(x), and taylorpoly6f(x). What do you notice?

2. Let $f(x) = \sin(x)$ and consider the Taylor Polynomials tnf about $a = \dfrac{\pi}{2}$. Suppose we want to approximate $\sin(2.1)$ by taylorpolynf(2.1).

   (a) Find a reasonable upper bound of $\left| f^{(n+1)}(w) \right|$ for w between $x = 2.1$ and $a = \dfrac{\pi}{a}$.

   (b) Find a reasonable upper bound of $\left| \dfrac{f^{(n+1)}(w)(x-a)^{n+1}}{(n+1)!} \right|$ for w between $x = 2.1$ and $a = \dfrac{\pi}{a}$.

   (c) Find a reasonable upper bound of $|\sin(2.1) - \text{t8f}(2.1)|$.

   (d) Plot $f(x)$ and taylorpoly8f(x) on $[1.1, 3.1]$.

   (e) How large must n be in order that $|\sin(2.1) - \text{taylorpolynf}(2.1)| < .00000000005$?

   (f) For that value of n, plot $f(x)$ and taylorpolynf(x) on $[1.1, 3.1]$.

3. Let $f(x) = x^6 + 2x^5 - 4x^3 + 3x^2 - 6x + 4$ and suppose we needed to make hand calculations of $f(x)$ for many values between .9 and 1.1.

   (a) By what reasoning might one decide to approximate $f(x)$ by $4(x-1)$?

   (b) How much error would result?

   (c) Plot $f(x)$ and $4(x-1)$ on $[0.9, 1.1]$.

   (d) Suppose, instead, you approximate $f(x)$ by $4(x-1) + 26(x-1)^2$. How much error would result?

   (e) Plot $f(x)$ and $4(x-1) + 26(x-1)^2$ on $[.9, 1.1]$.

4. Let $f(x) = \ln\left(\dfrac{1-x}{1+x} \right)$.

   (a) Find taylorpoly4f(x), where $a = 0$.

   (b) What error results when you use taylorpoly4f(x) to approximate $\ln(1.2)$?

5. Another way to approximate the error made in using the Taylor polynomial of degree n to approximate $f(x)$, is to calculate the area between $f(x)$ and taylorpolynf(x). In order to assure that a negative part will not cancel a positive part, we can square the difference between the two, and then take the square root of the final answer. This leads to the following formula for an approximation of the error between $f(x)$ and its n-th degree polynomial approximation taylorpolynf(x) over the interval [a,b].

$$\text{error}(a, b) = \sqrt{\int_a^b \left(f(x) - \text{taylorpolynf}(x)\right)^2 dx}$$

There are two directions that one can take in studying the behavior of the above error term. One approach is to study the error made in keeping n fixed but allowing the interval to change. This is the study you are asked to investigate in part (a). The other approach is to change the degree n of the Taylor polynomial but keep the interval of investigation fixed. This is the approach taken in part (b).

(a) Complete the entries in the following table. Assume that we are approximating f(x)=sin(x) by taylorpoly9fx) about x=0.

interval	error(a,b)
(-1,1)	_____
(-2,2)	_____
(-3,3)	_____
(-5,5)	_____
(-8,8)	_____
(-20,20)	_____

(b) Complete the entries in the following table. Assume the interval is fixed at $[-2\pi, 2\pi]$.

degree	error
3	_____
5	_____
8	_____
10	_____
15	_____

LAB # 24 — *Applications of Power Series*

---

1. Find the best rational approximation of the form $h(x) = \dfrac{a_0 + a_1 x + a_2 x^2 + a_3 x^3}{1 + b_1 x + b_2 x^2}$ for the

   function $f(x) = \dfrac{1}{1 - \tan(x)}$ on the interval $\left[\dfrac{-p}{2}, \dfrac{p}{5}\right]$. Be sure to compare the resulting Taylor

   polynomial and rational function to k(x). Which is the better approximation?

2. Find the best rational approximation of the form $h(x) = \dfrac{a_0 + a_1 x}{1 + b_1 x + b_2 x^2}$ for the

   function $f(x) = \dfrac{1}{\sin^2(x) - \cos^2(2x)}$ on the interval $\left[\dfrac{-\pi}{5}, \dfrac{\pi}{5}\right]$. Be sure to compare the resulting

   Taylor polynomial and rational function to k(x). Which is the better approximation?

3. Find a function y(x) that satisfies the differential equation y'''(x)-6y''(x)+11y'(x)-6y(x)=0 subject to the conditions y(0)=0, y'(0)=-2, and y"(0)=-8.

4. Find a function $y(x)$ that satisfies the differential equation $y''(x) + x\,y'(x) - 2y(x) = x - x^2$ subject to the conditions $y(0) = 1$ and $y'(0) = -2$.

5. Find a function $y(x)$ that satisfies the differential equation $y''(x) + x\,y(x) = \ln(1+x)$ subject to the conditions $y(0) = 1$ and $y'(0) = -1$.

LAB # 25 — *Parametric Equations and Quadratic Equations*

---

1. Consider the curve given by the parametric equation $p(t) = (\sin(t), t^{2/3})$. Plot this curve over the interval $[-\pi, \pi]$ and plot the tangent vectors for $t = -2, -1, 1,$ and 2.

2. A typical wave form that appears in physics is the curve $c(t) = (\sin(ht), \cos(kt))$. An obvious question to ask is "How would changing the parameter values h and k affect the shape of the curve?". Consider the shape of this wave form when $\dfrac{h}{k}$ is a rational number, such as $\dfrac{h}{k} = \dfrac{2}{5}$.

Then consider what happens when $\dfrac{h}{k}$ is irrational, such as $\dfrac{h}{k} = \dfrac{2}{\sqrt{3}}$. Use the

ParametricPlot command and plot the curve over a large interval.

3. The curve $-23x^2 - 2\sqrt{3}x^2 - 8xy - 14\sqrt{3}xy - 5y^2 - 2\sqrt{3}y^2 = 0$ represents a degenerate case. Identify the graph of this equation.

4. Consider $2x^2 - 3xy + 4y^2 - 3x + 4y - 2 = 0$.

Find:
(a) the **angle of rotation** that will eliminate the xy term;
(b) the **equivalent equation** for this curve that does not contain an xy term.
(c) the **class** to which the curve belongs.

5. A common task in Computer Aided Design (CAD) systems is that of drawing a curve segment between two given points that also has some additional properties. A typical approach to constructing such a curve is to give the two endpoints, p0 and p1, of the curved segment plus the beginning direction and the ending direction (the tangent vector at p0 and the tangent vector at p1). A parametric cubic curve, a parametric curve represented by a polynomial of third degree, leads to a very convenient way to generate a reasonable curve between p0 and p1.

The "geometric " form of such a curve is
p(u) =F1(u) p0 + F2(u) p1 + F3(u) tp0 + F4(u) tp1, where

$$F1(u) = 2u^3 - 3u^2 + 1, \quad F2(u) = -2u^3 + 3u^2, \quad F3(u) = u^3 - 2u^2 + u, \quad F4(u) = u^3 - u^2,$$

p0 = beginning point of the curve segment, **p1** = ending point of the curve segment,
tp0 = beginning direction of the curve segment, and **tp1** = ending direction of the curve segment.
In this exercise you are asked to study the effects accrued by changing the beginning direction, tp0. There are two different ways that this tangent vector can be changed, namely its direction relative to the positive x-axis and its magnitude. Throughout this experiment we will assume that p0 = (0,0), p1 = (1,1), and tp1 = {1,1}.

Use the following table of values to sketch the curves obtained by changing the vector tp0. Note that the first set of vectors changes only the direction angle of the tangent vector while the second set of values changes the magnitude of the tangent vector. To get a better understanding of how the curves are changing, label each curve as it is plotted and then use the **Show** command to simultaneously plot several of them on the same graph. The **GrayLevel** option can be used to distinguish between curves. Plot each curve over the interval $0 \le u \le 1$.

(a) Let tp0 = {cos(theta) , sin(theta) } where theta = 0, $\pi/3$, $\pi/4$, $\pi/2$, $5\pi/6$, and $5\pi/4$.

(b) Let tp0 = {r cos(theta) , r sin(theta) } where theta = $\pi/4$ and r = 1, 2,5,8,10, and 20.

Briefly state the charactristics exhibited by these curves as the indicated parameters are varied.

LAB # 26 — *Graphing in Polar Coordinates*

---

1. Plot the following curves for several values of the parameter. Then discuss the effect that changing the parameter has on the shape of the curve.

   (a) $r^2 = a^2 \cos(2t)$ (lemniscate of Bernoulli)

   (b) $r = e^{at}$ (logarithmic or equiangular spiral)

   (c) $r = \dfrac{a}{t}$ (hyperbolic or reciprocal spiral)

   (d) $r^2 = \dfrac{a^2}{t}$ (lituus)

   (e) $r^2 = a^2 \sin(2t)$ (two - leaved rose lemniscate)

2. Using the **Show** command, show the following curves on the same graph:

   $$r = \cos(t) + \sin^3(\tfrac{5}{2}t), \ \ 0 \le t \le 2\pi$$

   $$r = \frac{1}{3}\cos(3t), \ \ \frac{\pi}{6} \le t \le \frac{5\pi}{6}$$

   $$r = \frac{5}{4}\cos(7t), \ \ \frac{13\pi}{14} \le t \le \frac{15\pi}{14}$$

   $$r = -\frac{1}{6}\cos(3t), \ \ \frac{5\pi}{6} \le t \le \frac{7\pi}{6}$$

   (Use the **Axes- > None** option for the plotting.)

   Does this resemble a **butterfly**? See if you can improve the picture.

3. Example 1 concerned $r = m \sin(nt)$, $0 \le t \le 2\pi$. Prove the conjecture you made in **1.8**.

4. The following refer to the limaçon $r=1+k \cos(t)$ in EXAMPLE 3.
   (a) Are there shapes other than the three you have already seen? (Refer to 3.6.1, 3.6.2, and 3.6.3.) What values of k have been omitted?
   (b) In addition to the similarities you have already seen, see if you can find one or more points that are in common to all the graphs.

(c) Prove that your answers to 3.6.1, 3.6.2, and 3.6.3 are correct. (This may involve derivatives to help determine slopes exactly, and perhaps even second derivatives to determine concavity.) HINT. How many vertical tangents are there when the curve has a "dent"? How many vertical tangent lines are there when the curve has the shape of an "egg"?

(d) List other questions (and their answers) that came up during your investigation.

5. Consider **r=a+b sin(t)**. Conduct the same type investigation on these curves as was done on the curves r=a+b cos(t) in EXAMPLE 3.

LAB # 27 — *Vector Algebra*

---

1. Approximate the angle between the vectors $\mathbf{a} = \left\langle \frac{2}{5}, \frac{5}{3}, \frac{8}{9} \right\rangle$ and $\mathbf{b} = \left\langle \frac{8}{7}, \frac{3}{10}, \frac{7}{6} \right\rangle$.

2. (a) Find two vectors perpendicular to both $\left\langle \frac{83}{31}, \frac{-46}{53}, -1 \right\rangle$ and $\left\langle \frac{23}{30}, \frac{8}{5}, \frac{1}{7} \right\rangle$; and

   (b) Find two unit vectors perpendicular to both $\left\langle \frac{83}{31}, \frac{-46}{53}, -1 \right\rangle$ and $\left\langle \frac{23}{30}, \frac{8}{5}, \frac{1}{7} \right\rangle$.

3. Find all values of x so that each pair of vectors is perpendicular. If there are no values of x, state why not.

   (a) $\langle 1+2x, -10+2x, 7+8x \rangle$ and $\langle 4-4x, 5x, 5-8x \rangle$; and

   (b) $\langle 8-2x, 7+9x, 10-10x \rangle$ and $\langle 2-x, -3-7x, -6+4x \rangle$.

4. The angles between two curves at a point of intersection are the angles between the curves' tangents at that point. Find the angles between the point of intersection of the graphs of

   $f(x) = 1 + 2x + \sin(6x)$ and $g(x) = 2x^2$.

LAB # 28 — *Parametric Curves*

---

1. Consider the class of curves petal(k,t) = {cos(t) sin(kt) , sin(t) sin(kt)} obtained by varying the parameter k.

   (a) Graph the curve for k = 2, 3, 4, 5, 6, and 7, on the interval $[0, 2\pi]$.

   (b) How many petals would the curve contain if k = 8? If k = 9? If k = 101? If k = 102?

   (c) How is the shape of the curve changed if the value of k in just one term is changed? Graph each of the following curves and compare to that of the general form given.

       (i)   curve(t) = {cos(t) sin(4t) , sin(t) sin(3t) }

       (ii)   curve(t) = {cos(t) sin(4t) , sin(2t) sin(4t) }

       (iii)   curve(t) = {cos(2t) sin(4t) , sin(t) sin(4t) }

       (iv)   curve(t) = {cos(4t) sin(t) , sin(t) sin(4t)}

   (d) How does the shape of the following compare to that of the general form given?

       curve(t) = {cos(t) sin(3t) , sin(t) sin(4t) + cos(t) sin(t)}

2. Graph the curve given by the equation $curve(t) = \left\{ \dfrac{-3t(4-t^6)}{(t^8+1)}, \dfrac{-3t^2(4-t^6)}{(t^8+1)} \right\}$ on the interval $[-10, 10]$.

3. Consider the class of curves $c(t) = \{3\cos(t) + \cos(kt), 3\sin(t) + \sin(kt)\}$ obtained by varying the parameter k.
   (a) Graph the curve for k= 2, 3, 4, 5, 6, and 7, on the interval $[0, 2\pi]$.
   (b) Describe the general shape of the curve for each value of k.
   (c) Consider how the shape is changed if the value for k in just one term is changed. For example, plot the curve for k=4 except change the cos(4t) term to cos(5t).
   HINT. When you plot these curves, change the scale ratio for the x and y-axis to 1. This can be done by including the plot option **AspectRatio -> 1**

4. Consider the curves defined by $\begin{cases} x(k,t) = k\cos(t) + \cos(kt) \\ y(k,t) = k\sin(t) + \sin(kt) \end{cases}$. Plot the curve on $[0, 2\pi]$ for several values of k and compare the shapes, etc. to the corresponding curves in Example 1.

5. Consider the curves defined by $\begin{cases} x(k,t) = k\cos(t) - \cos(kt) \\ y(k,t) = k\sin(t) + \sin(kt) \end{cases}$. Plot the curve on $[0, 2\pi]$ for several values of k and compare the shapes, etc. to the corresponding curves in Example 1.

6. Consider the curves defined by $\begin{cases} x(k,t) = k\cos(t) - \cos(kt) \\ y(k,t) = k\sin(t) - \sin(kt) \end{cases}$. Plot the curve on $[0, 2\pi]$ for several values of k and compare the shapes, etc. to the corresponding curves in Example 1.

7. The rotary engine was first made somewhat well-known in the early 1950's by Felix Wankel. It features a rotor which moves around a cavity that in two-dimensions is shaped somewhat like a peanut. The shape of the "peanut" is defined by the parametric equations x=r cos(t)+b cos(3t) and y=r sin(t)+b sin(3t) where r and b are indicated by the figure below. (The outer circle rolls around the inner circle.)

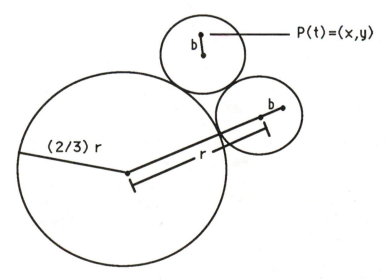

❏ EXERCISES ❏

(a) What is the radius of the outer circle?

(b) What restrictions are on the size of b?

(c) Using **ParametricPlot** with appropriate values of t and letting r=1, plot the "peanut" for the maximum value of b.

(d) Experiment with smaller values of b and state your conclusions about the effect on the shape of the peanut. Does your conclusion "make sense"? Explain.

LAB # 29 — *Surfaces In 3-Space*

For each of the functions f in Exercises 1-4,

(a) plot f over an appropriate region,

(b) obtain a contour plot of f over the same region and explain how it relates to the surface z=f(x,y),

(c) plot the curves obtained by slicing the surface $z = f(x,y)$ with 4 or 5 planes parallel to the xz-plane.

(d) plot the curves obtained by slicing the surface $z = f(x,y)$ with 4 or 5 planes parallel to the yz-plane.

1. $f(x,y) = y^2 \sin(x)$

2. $f(x,y) = \dfrac{x^2 y^2}{\sqrt{1 + x^6 + y^6}}$

3. $f(x,y) = \dfrac{xy}{1 + x^2 + y^2}$

4. $f(x,y) = \dfrac{xy}{1 + x^2 + y^4}$

5. Compare and contrast the functions and graphs in Exercises 2, 3, and 4.

6. $f(x,y) = \sin\left(\dfrac{2x^2 + y^2}{x^2 + y^2}\right)$

(a) This function oscillates a lot once it gets just a small distance away from the origin. Thus, you might want to plot this over rather small intervals about the origin, such as [-2.5,2.5]x[-2.5,2.5].

(b) Redraw the graph using the options **ViewPoint->{0,2,.3}** and **Boxed->False**

(c) What happens to f(x,y) for "large" values of x and y?".

7. One convenient way of representing a surface patch in 3-space is to use a parametric equation of the form p(u,w) = {x(u,w) , y(u,w) , z(u,w) } for a≤ u ≤ b , c≤w≤d. Thus p(u,w) is interpreted as a continuous deformation of the rectangle rect = [a,b]x[c,d]. The particular deformation, in this case, being described by the function p(u,w).

(a) Plot the surface patch given by the parametric equation p(u,v) = {sin(u) , cos(w) , u+w}, 0≤u≤2π , 0≤w≤2π by **Entering**
ParametricPlot3D[p[u,w], {u,0,2Pi}, {w,0,2Pi}, BoxRatios->{1.25,1.25,2}]

242

(b) A typical representation of a surface patch in CAD systems and computer graphics is a parametric cubic surface , that is, a surface defined by an equation of the form

$$p(u,w) = \sum_{i=0}^{3} \sum_{j=0}^{3} a_{ij} u^i w^j$$

Thus there are 16 coefficients that must be specified in order to be able to uniquely define this function. One natural set of input is

$p(0,0), p(0,1), p(1,0), p(1,1)$ - - - the four corner points

$p_u(0,0), p_u(0,1), p_u(1,0), p_u(1,1)$ - - - the beginning tangent directions at the four corner points

$p_w(0,0), p_w(0,1), p_w(1,0), p_w(1,1)$ - - - the tangent vectors in the direction w at the four corner points

$p_{uw}(0,0), p_{uw}(0,1), p_{uw}(1,0), p_{uw}(1,1)$ - - - the rate of change of the tangent vectors at the corner points

NOTE. The symbol $p_u(u_0, w_0)$ means $\dfrac{\partial p}{\partial u}$ evaluated at the point (u_0, w_0). Similarly.

$p_{uw}(u_0, w_0)$ means $\dfrac{\partial^2 p}{\partial u \partial w}$ evaluated at the point (u_0, w_0).

Having specified these 16 coefficients, the equation for p(u,w) is defined as

$$p(u,w) = [f1(u), f2(u), f3(u), f4(u)] \times \begin{bmatrix} p(0,0) & p(0,1) & p_w(0,0) & p_w(0,1) \\ p(1,0) & p(1,1) & p_w(1,0) & p_w(1,1) \\ p_u(0,0) & p_u(0,1) & p_{uw}(0,0) & p_{uw}(0,1) \\ p_u(1,0) & p_u(1,1) & p_{uw}(1,0) & p_{uw}(1,1) \end{bmatrix} \times \begin{bmatrix} f1(w) \\ f2(w) \\ f3(w) \\ f4(w) \end{bmatrix}$$

where $f1(u) = 2u^3 - 3u^2 + 1$, $f2(u) = -2u^3 + 3u^2$, $f3(u) = u^3 - 2u^2 + u$, and $f4(u) = u^3 - u^2$

```
f1[u_]:=2u^3-3u^2  +1
f2[u_]:=-2u^3 + 3u^2
f3[u_]:=u^3 -2u^2 + u
f4[u_]:=u^3 - u^2
F[u_]:={f1[u],f2[u],f3[u],f4[u]}

Clear[ p00, p01, p00w, p01w, p10, p11, p10w, p11w,
      p00u, p01u, p00uw, p01uw, p10u, p11u, p10uw, p11uw ]
B={{p00,p01,p00w,p01w},   {p10,p11,p10w,p11w},
     {p00u,p01u,p00uw,p01uw}, {p10u,p11u,p10uw,p11uw}}

B=F[u].B.Transpose[{F[w]}]

p00={0,0,10};   p01={10,0,0};
p10={0,10,8};   p11={18,10,0};
p00w={16,4,0};   p01w={0,-4,-16};
p10w={24,4,0};   p11w={0,-4,-14};
```

```
p00u={0,10,-8};   p01u={2,10,0};
p10u={0,10,-8};   p11u={2,10,0};
p00uw={8,8,0};    p01uw={-8,0,2};
p10uw={8,8,0};    p11uw={-8,0,2};

pts=Flatten[Simplify[B]]
```

Sketch the graph of p(u,v) by executing the following command.

ParametricPlot3D[pts, {u,0,1}, {w,0,1}, Boxed->False]

(c) By differentiating the function p(u,w), show that

$$\frac{\partial p}{\partial u}(0,0)=\text{p00u}, \quad \frac{\partial p}{\partial u}(1,0)=\text{p10u}, \quad \frac{\partial p}{\partial u}(0,1)=\text{p01u}, \quad \frac{\partial p}{\partial u}(1,1)=\text{p11u}.$$

LAB # 30 — *Partial Derivatives*

---

1. For each of the following functions,
   - (i)   define the function,
   - (ii)   find all critical points of f(x,y). (i.e. all points (x0,y0) for which fx=0 and fy=0, or either fx or fy does not exist).
   - (iii) for each critical point, use the command curveplot to graph the lift of the x-axis onto the surface z=f(x,y) and the lift of the y-axis onto the surface z=f(x,y).

   (a) $f(x,y)=e^{-(x^2-y^2)}, \quad -\frac{3}{2}\le x\le\frac{3}{2}, \quad -\frac{3}{2}\le y\le\frac{3}{2},$

   (b) $g(x,y)=\sin(x^2)\cos(y), \quad -\pi\le x\le\pi, \quad 0\le y\le 2\pi,$

   (c) $h(x,y)=\dfrac{3xy}{x^2+y^2}, \quad -2\le x\le 2, \quad -2\le y\le 2, \ (x,y)\ne(0,0),$

   (d) $k(x,y)=x^4+y^4-4xy.$

2. When a pollutant such as nitric oxide is emitted from a smokestack of height h meters, the long-range concentration C(x,y) of the pollutant at a point x kilometers from the smokestack at a height of y meters can often be represented by $C(x,y)=\dfrac{a}{x^2}\left[e^{\dfrac{-b(y-h)^2}{x^2}}+e^{\dfrac{-b(y+h)^2}{x^2}}\right],$

   where a and b are positive constants that depend on atmospheric conditions and the pollution emission rate.

   Suppose that $a=200$, $b=0.02$, and $h=10$. Compute and interpret $\dfrac{\partial C}{\partial x}$ and $\dfrac{\partial C}{\partial y}$ at the point $(2,5)$.

3. Laplace's equation for the function $f(x,y)$ is $\dfrac{\partial^2 f}{\partial x^2} + \dfrac{\partial^2 f}{\partial y^2} = 0$.

Show that the following functions satisfy Laplace's equation.

(a) $f(x,y) = e^x \sin(y)$.

(b) $f(x,y) = e^{-x} \cos(y)$

(c) $f(x,y) = \ln(\sqrt{x^2 + y^2})$.

4. Find the equation of the tangent plane for each of the following equations at the specified point.

(a) $f(x,y) = x^2 + y^2 - 4x + 2y$ at $(x0, y0) = (2,1)$,

(b) $g(x,y) = \sqrt{9 - x^2 + 3y^2}$ at $(x0, y0) = (2,1)$,

(c) $h(x,y) = e^{-x} \sin(y)$ at $(x0, y0) = \left(1, \dfrac{\pi}{6}\right)$.

LAB # 31 — *Directional Derivatives*

---

1. Let $f(x,y) = \dfrac{x^2 y + x y^2}{10}$.

(a) Find the direction derivative, Dv, of f at **P** = (2,1) and $\theta = \pi/3$.
(b) What is the equation of the the curve vcurve(t)?
(c) What is the equation of the tangent line to vcurve(t) at **P**?
(d) Use tanplot to plot the surface, the curve vcurve(t), and the tangent line. Use {x,-3,3} and {y,0,3} to specify the region over which to plot the surface z = f(x,y).

2. The surface of a boat is represented by a region D in the xy-plane such that the depth (in feet)

under the point (x,y) is given by the equation $depth(x,y) = 300 - 2x^2 - 3y^2 + xy$.

(a) In what direction should a boat at **P**(4,9) move in order for the depth of the water to decrease most rapidly?
(b) In what direction does the depth remain the same?

3. A metal plate is located in the xy-plane such that the temperature T at (x,y) is inversely proportional to the distance from the origin. If the temperature at P(2,3) is 100 degrees Farenheit, then
(a) What is the rate of change of T at P in the direction of **i** + **j**?
(b) In what direction does T increase most rapidly?
(c) In what direction does T decrease most rapidly?
(d) In what direction will the change in temperature be the same?

4. Let $f(x,y) = \begin{cases} \dfrac{xy}{x^2 - y} & (x,y) \neq (0,0) \\ 0 & (x,y) = (0,0) \end{cases}$.

   (a) Find the directional derivative of f at $\mathbf{P} = (0,0)$ for each of the following directions.

      (i) $\theta = 0$

      (ii) $\theta = \pi/3$

      (iii) $\theta = \pi/4$

      (iv) $\theta = \pi/2$

      (v) $\theta = \pi$.

      HINT. You will have to use the limiting procedure to find Dv.

   (b) Does f have a tangent plane at $\mathbf{P}$? Why or why not?

   (c) Is f continuous at $\mathbf{P}$? Why or why not?

   (d) Use the **Plot3D** command to plot f in a region about $\mathbf{P}$.

LAB # 32 — *Critical Points*

1. Let $f(x,y) = -x^3 + 2y^3 + 27x - 24y + 3$.

Classify each critical point as either a local maximum point, a local minimum point, or a saddle point.

2. Let $f(x,y) = x^4 + y^4 - 50xy^2 - 50x^2y - 70y^2 - 15x^2 + 64x$.

Classify each critical point as either a local maximum point, a local minimum point, or a saddle point.

3. Let $f(x,y) = \dfrac{4y + x^2 y^2 + 8x}{xy}$.

   (a) What points must be excluded from the domain of f?

   (b) Classify each critical point as either a local maximum point, a local minimum point, or a saddle point.

   (c) For each critical point, construct a contour plot restricting the region in the xy-plane so as to include only a single critical point.

   (d) Plot f over the region $-6 \leq x \leq 6$, $-6 \leq y \leq 6$.

   (e) Plot f in a neighborhood of each critical point. Restrict the region in the xy-plane over which f is being plotted to a region that does not include any points of the plane that are not in the domain of f.

4. Let $f(x,y) = \dfrac{\sin(x^2 + y^2)}{x^2 + y^2 + 1}$.

Classify all critical points that lie in the circular region $C = \{(x,y) \mid x^2 + y^2 \leq 10\}$.

HINT. Plot f over the above region. Include the option **PlotPoints -> 30** to obtain a better view of the graph. It might be better to use some known trigonometric identities in analyzing the set of critical points rather than just relying on the calculus techniques.

5. Let $f(x,y) = \dfrac{-2x^3}{3} - \dfrac{x^4}{4} + \dfrac{2x^5}{5} + \dfrac{x^6}{6} + xy^2$.

   Classify each critical point as either a local maximum point, a local minimum point, or a saddle point.

6. Let $f(x,y) = xy\, e^{-\left(x^2+y^2\right)}$

   (a) Plot f over the region $-3 \le x \le 3$, $-3 \le y \le 3$.
   (b) Classify each critical point as either a local maximum point, a local minimum point, or a saddle point.
   (c) Construct a contour plot of f over the region $-5 \le x \le 5$, $-5 \le y \le 5$.

LAB # 33 — *Applied Max/Min Problems (several variables)*

---

1. Find the point on the plane $2x + 3y + z - 12 = 0$ that is nearest the origin.

2. A manufacturer makes two models of an item, a standard and a deluxe. It costs $50 to manufacture the standard model and $72 for the deluxe. A market research firm estimates that if the standard model is priced at x dollars and the deluxe at y dollars, then the manufacturer will sell $500(y - x)$ of the standard items and $10{,}000 + 500(x - 2y)$ of the deluxe each year. How should the items be priced to maximize the profit?

3. A cable line is to be built connecting point A and B and must pass through regions where the construction costs differ. See the sketch on the next page for an outline of the region over which the cable must be laid. If the cost per mile in dollars is 3k from A to C, 2k from C to D, and k from D to B, find the values of x and y that will result in the minimum cost.

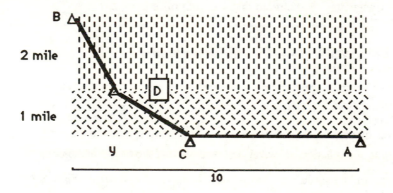

4. **Approximation of data by a quadratic curve.** We can use the same approach that we used in Example 3 to determine the "best" fitting quadratic, the curve $y = ax^2 + bx + c$, that "best" fits a collection of data points in the sense that the mean square error,

$$\text{error}(a, b, c) = \sum_{i=1}^{n} (y_i \text{(on the quadratic)} - y_i \text{(data point)})^2, \text{ is minimized.}$$

(a) Determine $\text{error}_a \left(= \dfrac{\partial \, \text{error}}{\partial a} \right)$, error_b, and error_c.

(b) By labeling the sums that appear in the equations for error_a, error_b, and error_c, similar to what we did in Example 3, determine the minimizing values for the coefficients a, b, and c.

(c) To check out the formula that we just derived, lets construct a "noisy" collection of data from a quadratic equation. To do this, let $y = f[\,x\,] = 2x^2 - 3x + 4$. The following commnads will generate a set of noisy data.

x = Table[i , {i , -2, 2 , .25}]
y= Table[f[i]+ Random[Real , {-1 , 1}] , {i , -2 , 2 , .25}]

Using this set of data, calculate the quadratic coefficients a,b, and c. Then , similar to what we did in Example 3, plot both the data points and the approximating curve on the same graph. Show this graph.

(d) Using the set of data generated above, calculate the "best" approximating line using the formulas developed in Example 3. Also plot the data points, the approximating quadratic curve, and the approximating line on the same graph.

5. The computer center of an exploration team at the North Pole is in the shape of a rectangular box (parallelepiped). Its volume is 6000 cubic feet. In a given period of time, the heat loss through the ceiling is three times greater than the loss through the floor and twice as great as the loss through the walls. Find the dimensions of the room that will minimize heat loss.

LAB # 34 — *Lagrange Multipliers*

---

1. Find the extrema of the function $f(x,y) = x^3 + 2y^3$ subject to the constraint $x^2 + y^2 = 1$.

2. A nuclear explosion is detonated in space and the radiation cloud spreads in the form of a sphere with center at the point of explosion. If we choose axes so that the origin is the point of explosion, what points on the surface $x^2 - yz = 1$ will be the first to receive radiation?

3. Consider a tin can in the shape of a right circular cylinder. (The can has a top and a bottom.)
   (a) Find the dimensions of the can that will maximize the volume if the surface area is S square centimeters.

(b) Find the dimensions of the can that will minimize the surface area if the volume is V cubic centimeters.

4. Find the dimensions of a rectangle that will maximize the area if the perimeter is P meters.

5. Find the dimensions of a rectangle that will minimize the perimeter if the area is A square meters.

6. The computer center of an exploration team at the North Pole is in the shape of a rectangular box (parallelepiped). Its volume is 6000 cubic feet. In a given period of time, the heat loss through the ceiling is three times greater than the loss through the floor and twice as great as the loss through the walls. Find the dimensions of the room that will minimize heat loss. (This problem was also in the previous set of exercises.)

LAB # 35 — *Double Integrals*

---

1. Evaluate the iterated integral $\int_0^1 \int_0^y x\sqrt{y^2 - x^2}\, dx\, dy$. What does it represent? Be specific.

2. Evaluate the iterated integral $\int_0^5 \int_0^{\sqrt{25-x}} \sqrt{25 - x^2 - y^2}\, dy\, dx$. What does it represent? Be specific.

3. The iterated integral $\int_{-8}^8 \int_{-\sqrt{64-x^2}}^{\sqrt{64-x^2}} \left(16 - 4x^2 - 4y^2\right) dx\, dy$ corresponds to the volume of

   the region V below the graph of $z = 16 - 4x^2 - 4y^2$ and above the region in the xy – plane

   bounded by the graphs of $x = -8$, $x = 8$, $y = \sqrt{64 - x^2}$ and $y = -\sqrt{64 - x^2}$. (See the graph

   below.) Find V.

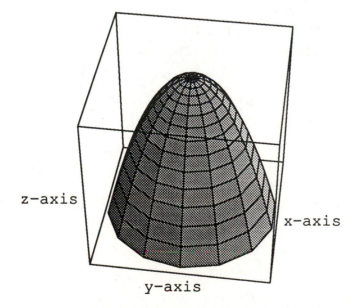

4. Let $f(x,y) = 4 - x^2 - y^2$. Then the iterated integral

$$\int_{-2}^{2} \int_{-\sqrt{4-x^2}}^{\sqrt{4-x^2}} \sqrt{\left(f_x(x,y)\right)^2 + \left(f_y(x,y)\right)^2 + 1} \; dy \, dx \text{ corresponds to the surface area of } f(x,y)$$

above the xy - plane. Calculate the surface area of $f(x,y)$ above the xy - plane.

LAB # 36 — *Triple Integrals*

---

1. Find the value of each of the following integrals.

   (a) $\displaystyle\int_{0}^{6}\int_{0}^{4-2x/3}\int_{0}^{3-x/2-3y/4} (x + y\,z) \; dz \, dy \, dx$

   (b) $\displaystyle\iiint_{Q} x\,y\,e^{z} \; dV$ where Q is the region bounded by the graphs of $y = x^2$, $z = 4 - y$, and $z = 0$.

2. Find the volume of the solid formed by the intersection of the cone $x = \sqrt{y^2 + z^2}$ and the paraboloid $x = 6 - y^2 - z^2$.

   HINT. You might want to plot both these solids separately and then merge the graphs together. How would you plot the skeleton outline of this solid?

3. Using the procedure in Example 2 as a guide, state all steps in obtaining the outline of the solid shown on the previous page.

4. Sketch the region in the first octant that is bounded by the graphs of $y = x^3$, $y = 8$, and $z = 4$. Also find its volume.

5. Find the center of mass of the solid bounded by the graphs of $y = x^3$, $y = x$, and $z = x + 2$.. (Assume that the density at the point is directly proportional to the distance from the xz - plane.)

6. Find the moment of inertia of the solid located in the first octant that is bounded by the graphs of $y = x^3$, $y = 8$, and $z = 4$. (Assume that the density at a point P is directly proportional to its distance from the xy - plane.)

7. A cross-section of an experimental airfoil is the lamina shown on the top of the next page. The arc ABC is elliptical, whereas the two arcs AD and CD are parabolic. Find its center of mass if the density at any point is 100 plus the distance from the origin.

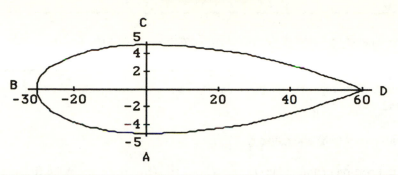

LAB # 37 — *More on Volumes and Triple Integrals*

1. Find the volume of the solid bounded by the plane $z = 4$ and the cone $z = \sqrt{x^2 + y^2}$.

2. Find the volume of the solid bounded by the graphs of $z = x^2 + y^2$ and $z = \dfrac{y^2}{4} - \dfrac{x^2}{9} + 4$.

3. Find the volume of the solid bounded by the graphs of $z = \sqrt{x^2 + y^2}$ and $z = 4 - x^2 - y^2$.

4. A bearing sleeve is formed as follows. First construct a pipe formed by the graphs of $r = 1$, $r = 4$, $z = 0$, and $z = 9$. Secondly, the sleeve is constructed by hollowing out the inside of the pipe with a solid in the shape of the paraboloid $z = x^2 + y^2$. A cross-sectional view of the bearing sleeve is shown below.

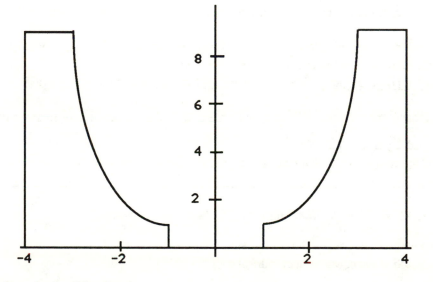

Find the volume of the sleeve.

251

LAB # 38 — *Surface Area & Other Applications of Multiple Integrals*

---

1. Find the surface area of the region of the plane $2x+4y+2z=3$ contained within the region bounded by
   (i) $x=0$, $y=0$, and $z=0$;
   (ii) the unit circle; and
   (iii) the circle with center (a,b) and radius r.

2. Find the area of the portion of the surface (a) $z = x + y^2$ and (b) $z = y + x^2$ in the first octant

   bounded by the plane (i) $x + y = 1$; (ii) $x + y = m$; and (iii) $\dfrac{x}{a} + \dfrac{y}{b} = 1, a, b > 0$.

3. Find the value of the integral $\displaystyle\iiint_Q \delta(x,y,z)\,dV,$ where $\delta(x,y,z) = x\,y\,e^z$ and Q is the region

   bounded by the graphs of $y = x^2$, $z = 4 - y$, and $z = 0$.

4. (a) Use the Monte Carlo method with $n = 100$ to estimate the mass of the solid with shape Q,

   where Q is the region bounded by the graphs of $y = x^2$, $y = x$, $z = y + 2$, and $z = 0$ if the
   density at a point P is directly proportional to the distance from the xy - plane.

   (b) How much would your answer differ if you used $n = 200$? $n=300$?

   (c) How much change would result from the density at a point P being directly proportional to the
   distance from the origin instead of the distance from the xy-plane?

5. The volume of a complicated region Q in 3-space can be estimated by the Monte Carlo method as
   follows:
   1. Determine a "nice" region R having known volume and containing Q. (e.g. a parallelepiped),
   2. Generate n random points in R and count those also in Q.
   3. If m of the n points are in Q, then the volume of Q is approximately m/n that of R.

   Using the Monte Carlo method, estimate the volume of the following region for $n=100$ and 200.

   $-1 \le x \le 1, \ -1 \le y \le 1, \ -1 \le z \le 1$

   $x^2 + \sin(y) \le z, \ x - z + e^y \le 1$

LAB # 39 — *Cylindrical and Spherical Coordinates*

---

1. Describe the graph of the equation in cylindrical coordinates $r=\csc(\theta)+\sec(\theta)$.

2. Find the volume of the solid bounded by the plane $z = 4$ and the cone $z = \sqrt{x^2 + y^2}$.

3. Change the equation with cartesian coordinates $4x-3y+z=12$ to spherical coordinates.

❑ EXERCISES ❑

4. In the experimental testing of the effects of exposure to a radioactive gas, a chamber in the shape of a paraboloid $z = 36 - x^2 - y^2$ is used. Assume $z \geq 0$, units in centimeters, and that the density of radioactive energy is to be $(5)\left(10^{-10}\right)$ joules per cubic centimeter. Find the total amount of radioactive energy required in the chamber.

5. Using **CylindricalPlot3D**, graph the following:

(a) $z = 4r^2$, $0 \leq r \leq 1$, $0 \leq \theta \leq 2\pi$.

(b) $z = \sqrt{16 - r^2}$, $0 \leq r \leq 1$, $0 \leq \theta \leq 2\pi$.

Glossary

Abs
Abs[x] computes the absolute value of x.

Apart
Apart[expression] computes and returns the partial fraction decomposition of **expression**.

Apply
Apply[Plus,listofnumbers] computes the sum of **listofnumbers**. More generally, **Apply[f,list]** computes **f[list]**.

AspectRatio
AspectRatio is an option used with graphics commands like **Plot**, **ParametricPlot**, and **Show** which specifies the ratio of the height to the width of the displayed graphics image. The default value is **AspectRatio->1/GoldenRatio**, where **GoldenRatio** represents the built-in constant $\left(1+\sqrt{5}\right)/2$.

Axes
Axes is an option used with graphics commands which specifies whether axes are displayed in the graphics image or not. **Axes->True** specifies that axes are displayed; **Axes->False** specifies that no axes are displayed.

Boxed
Boxed is an option used with three-dimensional graphics commands like **Plot3D**. The option **Boxed->False** indicates that no box is to be drawn around the displayed graphics; **Boxed->True** indicates that a box is to be drawn around the displayed graphics.

BoxRatios
BoxRatios is an option used with three-dimensional graphics commands like **Plot3D**. The option **BoxRatios->{xn,yn,zn}** specifies that the ratio of the length of the x-axis to the length of the y-axis to the length of the z-axis is **xn:yn:zn**.

Cancel
Cancel[expression] cancels common factors from the numerator and denominator of **expression**.

Chop
Chop[expression] returns the expression obtained by replacing real numbers in **expression** close to zero by zero.

Clear
Clear[symbol] clears definitions of **symbol**.

ColumnForm
ColumnForm[array] displays array in a traditional row-and-column form.

ContourPlot
ContourPlot[f[x,y],{x,xmin,xmax}, {y,ymin,ymax}] produces a graph of several contour levels for the function of two variables **f[x,y]** over the region [xmin,xmax] $\times$ [ymin,ymax].

Contours
Contours is an option used with **ContourPlot**. **Contours->n** specifies that **n** equally spaced contours are to be used to create the graph.

Coordinates
Coordinates is a command contained in the package **VectorAnalysis.m** and loaded with the command **<<Calculus`VectorAnalysis`**. **Coordinates[system]** returns the coordinate variables in the coordinate system **system**.

Cos
Cos[x] computes the cosine of x.

curvetnplot
curvetnplot[f,{x0,y0},thetaset] will plot the curves and their tangent lines determined by restricting f to lie along the lines in the xy-plane whose angles of inclination are specified in the set thetaset. (This is not a standard command.)

D
D[f[x],x] computes the derivative of **f[x]** with respect to x; **D[f[x],{x,n}]** computes the nth derivative of **f[x]** with respect to n; and **D[f[x,y],x,y]** yields the partial derivative of **f[x,y]** with respect to y then x.

delnbd
s=delnbd[{a,b},pt,n] creates a set s of n random numbers between a and b, none of which are equal to the number x=pt. Thus **delnbd** returns a list of n random numbers from the deleted neighborhood **(a,b,)-{pt}** of pt. (This is not a standard command.)

Denominator
Denominator[fraction] returns the denominator of **fraction**.

DensityPlot
DensityPlot[f[x,y],{x,xmin,xmax}, {y,ymin,ymax}]] makes a density plot of f(x,y) on the rectangle [xmin,xmax] × [ymin,ymax].

DisplayFunction
DisplayFunction is an option for graphics functions like **Show**, **Plot**, and **Plot3D** which indicates how the resulting graphics object is displayed. The default setting, **DisplayFunction->$DisplayFunction,** indicates that the graphics objects produced are displayed; **DisplayFunction->Identity** generates no display.

Do
Do[expression[i],{i,imin,imax,istep}] computes **expression[i]** for values of the variable **i** from **i=imin** to **i=imax** in steps of **istep**. If **istep** is not included, the default stepsize is 1.

Exp
Exp[x] represents the exponential function e^x.

Expand
Expand[expression] carries out the multiplication of products and positive integer powers in **expression** so that the resulting output is expressed as a sum of terms.

ExpandDenominator
ExpandDenominator[expression] carries out the multiplication of products and positive integer powers in the denominator of **expression**.

ExpandNumerator
ExpandNumerator[expression] carries out the multiplication of products and positive integer powers in the numerator of **expression**.

Factor
Factor[poly] factors the polynomial poly over the integers.

FindRoot
FindRoot is used to approximate the numerical solution of an equation or a system of equations. This command is entered as **FindRoot[f[x]==g[x],{x,x0}]** where **x0** is the initial guess of the solution to the equation f(x)=g(x).

Flatten
Flatten[list] flattens **list** by removing nested parentheses.

Gamma
Gamma[x] represents the Gamma function $\Gamma(x) = \int_0^\infty t^{x-1}e^{-t}dt$.

generatepts
generatepts[s,n,a,b] is used to create a set **s** of **n** random points (x,y,z) in the region **Q**. The values x, y, and z are generated in the interval [a,b]. This command is used in conjunction with the command **ptinregion**. (This is not a standard command.)

Graphics
Graphics[primitives,options] denotes the graphical image of a two-dimensional object. These objects are viewed with **Show**. The primitives which may be used include **Circle**, **Point**, **Polygon**, **Rectangle**, and **Line** while the lengthy list of options includes **AspectRatio**, **Axes**, **DisplayFunction**, **Frame**, **PlotRange**, and **Ticks**. Graphics directives such as **GrayLevel**, **RGBColor**, and **Thickness** may also be included in a **Graphics** command.

GrayLevel
GrayLevel[intensity] is a directive used with graphics functions to indicate the level of gray to be used when displaying graphics objects. The value of intensity is between 0.0, representing black, and 1.0 , representing white.

Infinity
Infinity represents $+\infty$.

Integrate
Integrate is used to compute the exact value of definite and indefinite integrals. Integrals may be single or multiple: **Integrate[f[x],x]** attempts to determine the indefinite integral $\int f(x)dx$, **Integrate[f[x],{x,x1,x2}]** attempts to evaluate $\int_{x_1}^{x_2} f(x)dx$, and **Integrate[f[x,y],{x,x1,x2},{y,y1,y2}]** attempts to evaluate $\int_{x_1}^{x_2}\int_{y_1}^{y_2} f(x,y)dy\,dx$. **Integrate** is usually most useful in evaluating those integrals which depend on exponential, logarithmic, trigonometric, and inverse trigonometric functions if the result also involves these functions.

InterpolatingPolynomial

InterpolatingPolynomial[listofpairs, variable] returns a polynomial in **variable** which passes through each of the points specified in **listofpairs**.

inversegraph

inversegraph[f[x],{x,a,b}] plots **f(x), the inverse graph of f(x)**(in gray), and the line **y=x** (dashed). (This is not a standard command.)

Limit

Limit[expression,x->x0] attempts to determines the limit of **expression** as **x** approaches **x0**.

Line

Line[listofpoints] is the graphics primitive which represents the collection of line segments connecting consecutive points in the list of ordered pairs **listofpoints**.

linearfit

linearfit[list of points,m,n] joins each successive pair of points in the list, from point m to point n, with a line segment and then displays the graph. (This is not a standard command.)

ListPlot

ListPlot[{f1,f2,...}] plots the list of points {1, **f1**}, {2, **f2**}, ... ; ListPlot[{x1,y1},{x2,y2},...] plots the list of points {**x1,y1**}, {**x2,y2**}, ...Since the plot which results is a **Graphics** object, **ListPlot** has the same available options as **Graphics**. In addition, including the option **PlotJoined->True** causes the consecutive points in the graph to be connected by line segments.

Log

Log[x] represents the natural logarithm (base e) of x.

LogicalExpand

LogicalExpand[expression] expands out expressions in **expression** which involve the logical connectives **& &** and ‖ by applying appropriate distributive properties.

Map

Map is used with functions and lists to apply a function to particular elements of a list. **Map[f,list]** computes the value of **f** for each member of **list**.

Max

Max[listofnumbers] returns the maximum of the list of numbers **listofnumbers**.

Min

Min[listofnumbers] returns the minimum of the list of numbers **listofnumbers**.

N

N[expression] yields the numerical value of expression. **N[expression,n]** attempts to approximate **expression** to n digits of precision. With this command *Mathematica* performs computations with n-digit precision numbers. However, results may involve fewer than n digits.

Nest

Nest[f,x,n] computes the composition of f(x) with itself n times.

NIntegrate

NIntegrate[f,{x,x0,x0}] attempts to numerically compute $\int_{x_0}^{x_1} f(x)dx$.

Normal

Normal[expression] changes **expression** to a normal expression. If **expression** is a power series, then this is accomplished by removing the "big O" term which represents the omitted higher order terms of the series.

NRoots

NRoots[poly==0,variable] approximates the solutions of the polynomial equation **poly**=0, where **poly** represents a polynomial in x.

Numerator

Numerator[expression] returns the numerator of **expression**.

ParametricPlot

ParametricPlot is used to plot parametric curves in two dimensions.
ParametricPlot[{x[t],y[t]},{t,t0,t1}] graphs the curve given by x=x[t] and y=y[t] from t = t_0 to t = t_1.

ParametricPlot3D

ParametricPlot3D[{x[t,u],y[t,u],z[t,u]}, {t,tmin,tmax},{u,umin,umax}] graphs the surface defined by x=x(t,u), y=y(t,u), z=z(t,u) for $t_0 < t < t_1$ and $u_0 < u < u_1$.

paramplot[{x,y},{tstart,tfinish},tset]

plots the points (x[t],y[t]) for t in the interval [tstart,tfinish]. In addition, for each set tset = {t_1,t_2,....,t_n} of t values, the command will plot the tangent vector at each point (x(t_k),y(t_k)) for k = 1,2,...n. The tangent vector points in the direction of the curve. (This is not a standard command.)

Part ([[...]])

Part[expr,i] or **expr[[i]]** yields the ith part of **expression**.

Plot

Plot is used to plot functions of a single variable. **Plot[f[x],{x,xmin,xmax}]** graphs the function **f[x]** on the interval [**xmin,xmax**]. **Plot[{f1[x],f2[x],...},{x,xmin,xmax}]** graphs the functions **f1**, **f2**, ... simultaneously. The options used with **Graphics** are also available to **Plot** along with several more. Additional options as well as their default setting are **Complied->True, MaxBend->10, PlotDivision->20, PlotPoints->25,** and **PlotStyle->Automatic**.

plotptdiscontinuity

plotptdiscontinuity[f[x],{x,a,b},c] graphs the function **f[x]** with a single discontinuity at x=c on the interval [**a,b**]. (This is not a standard command.)

PlotPoints

PlotPoints is an option used with graphics commands like **Plot** and **Plot3D** which determines the total number of equally spaced sample points to be used. **PlotPoints->n** indicates that **n** points be used while with two variables **PlotPoints->n** implies that **n** points be selected in the direction of both coordinates; if different numbers are to be used in the two directions, then **PlotPoints->{nx,ny}** is used.

PlotRange

PlotRange is an option used with graphics functions like **Plot** to indicate the function values to be included in the graph. The setting **PlotRange->All** indicates that all points are to be included, **PlotRange->Automatic** implies that outlying points be omitted, and the setting **PlotRange->{min,max}** places definite limits on the y-values to be displayed in the graph.

PlotStyle

PlotStyle is an option used with **Plot** and **ListPlot** to indicate the manner in which lines or points are displayed. The setting **PlotStyle->style** implies that all lines or points be drawn using the graphics directive **style**. Such directives include **RGBColor**, **Thickness**, **Hue**, and **GrayLevel**. **PlotStyle** is often useful in multiple graphs.

Plot3D

Plot3D is used to graph functions of two variables. **Plot3D[f[x,y],{x,xmin,xmax},{y,ymin,ymax}]** yields a three-dimensional graph of the function **f[x,y]** over the region defined by [**xmin,xmax**]× [**ymin,ymax**] in the xy-plane.

Point

Point[coords] is the graphics primitive which represents a point with coordinates **coords** in two or three dimensions. The coordinates given in **coords** can be expressed as {x,y} or {x,y,z}. Points are rendered as circular regions and may be shaded or colored using **GrayLevel**, **Hue**, **RGBColor**, or **CMYKColor**. The size of the point is controlled by the option **PointSize**.

PointSize

PointSize[r] indicates that all **Point** elements be drawn as circles of radius **r** in the graphics object. The radius **r** is measured as a fraction of the width of the entire graph. The default setting in two dimensions is **PointSize[0.008]** while it is **PointSize[0.01]** for three dimensions.

PolarPlot

PolarPlot is contained in the package **Graphics.m** located in the **Graphics** folder and loaded with the command **<<Graphics`Graphics`**. **PolarPlot[r[theta],{theta,thetamin,thetamax}]** creates a polar graph of **r[theta]** for **thetamin ≤ theta ≤ thetamax**.

ptinregion

ptinregion[x,y,z] will return the value **TRUE** or **FALSE** depending upon whether or not the point (x,y,z) is in the region **Q**. This command is used to actually define the region **Q**. (This is not a standard command.)

randompoly

randompoly[deg] generates a polynomial p(x) with degree **deg** and random integer coefficients between -100 and 100. **randompoly[deg,n]** generates a polynomial p(x) with degree **deg** and random integer coefficients between **-n** and **n**. (This is not a standard command.)

randompolynomial
  randompolynomial generates a random polynomial p(x) with degree less than 50 and integer coefficients between -100 and 100. The definition for **randompolynomial** is in the initialization cell. (This is not a standard command.)

randomtriangle
  randomtriangle[length1, length2] shows a triangle with two sides of lengths **length1** and **length2**, and third side of random length between **length1 + length2** and │**length1-length2**│, and then uses Heron's Formula to compute its area. (This is not a standard command.)

ReplaceAll (/.)
  ReplaceAll[exp,rules], symbolized by **exp/.rules**, is used to apply a rule or list of rules, **rules**, to every part of an expression **exp**.

riemannsum
  riemannsum[f[x],{x,a,b},n,pt,graph] computes a Riemann sum for **f** by dividing **[a,b]** into **n** equal subintervals, picking **pt** inside each subinterval and computing the Riemann sum. **pt** may be chosen as **left, right, random,** or **any number between 0 and 1. (If left(right)** is chosen, the point taken in each subinterval is the **left(right)**-hand endpoint of the subinterval. If **random** is chosen, a point in each subinterval is chosen randomly. If **pt** is chosen as a number between 0 and 1, then the point chosen in each subinterval is the point that "fraction" of the distance from the left-hand endpoint to the right-hand endpoint.) If **graph** is not included, no graph is drawn. (This is not a standard command.)

secantline
  secantline[f[x],{x,a,b},c] draws the graph of f on the interval [a,b], a dashed graph of the line tangent to the graph of f at the point (c,f(c)) and a sequence of secant lines about x=c. **secantline[f[x],{x,a,b},c,tan->no]** draws the graph of f on the interval [a,b] and a sequence of secant lines about x=c. This version should be used when the curve f **does not have** a tangent line at (c,f(c)). (This is not a standard command.)

Series
  Series[f[x],{x,a,n}] computes the power series expansion for the function f[x] about x=a,

$$\sum_{n=0}^{\infty} \frac{f^{(n)}(a)}{n!} (x-a)^n \text{ , up to order } (x-a)^n.$$

Show
  Show is used to display two- and three-dimensional graphics objects. The graphics objects **graphics1, graphics2, ...** can be shown at once with **Show[graphics1,graphics2,...]**. Note that all options included in the **Show** command override options used in the original graphics commands. The option **DisplayFunction** specifies whether graphs are to be shown or not. The setting **DisplayFunction->$DisplayFunction** causes the graph to be shown while **DisplayFunction->Identity** causes the display to be suppressed.

Simplify
  Simplify[expression] attempts to determine the simplest form of **expression** through a sequence of algebraic transformations. In many cases, commands like **Expand** and **Factor** are useful prior to application of **Simplify**.

Sin
  Sin[x] represents the trigonometric sine function of x.

solidrev
  solidrev[f[x],{x,a,b},axis] yields a three-dimensional meshed image of the function f(x) defined on the domain [a,b] revolved about the x-axis or y-axis. **solidrev[f[x],{x,a,b},axis,solid]** yields a solid surface. The interval [a,b] is automatically divided into 10 subintervals. This may be changed by substituting **{x,a,b},n** for **{x,a,b}** where n is the desired number of subintervals. (This is not a standard command.)

Solve
  Solve[f[x]==g[x]] attempts to solve the equation f(x)=g(x) for x, as long as x is the only unknown in the equation; **Solve[f[x]==g[x],x]** attempts to solve the equation f(x)=g(x) for x. A system of equations may be solved with **Solve[{lhs1==rhs1,lhs2==rhs2,..}, {var1,var2,...}]** which solves the given equations for the variables listed.

Sqrt
  Sqrt[x] computes $\sqrt{x}$.

Sum

Sum[a[i],{i,n,m}] computes the finite sum
$$\sum_{i=n}^{i=m} a(i) .$$

sumf

sumf[f,s] finds the sum of the f(x,y,z)'s such that (x,y,z) is in the finite set s. (This is not a standard command.)

Table

Table[expression,{i}] generates n copies of expression. Table[expression[i],{i,imax}] creates a list of the values of expression from i = 1 to i = imax. A minimum value of i other than 1 is indicated with Table[expression,{i,imin,imax}] while a step size other than one unit is defined with Table[expression,{i,imin,imax,istep}].

TableForm

TableForm[list] prints list in a traditional row-and-column format in the form of a rectangular array of cells.

tanplot

tanplot[f, theta, {x0,y0}, {x,a,b}, {y,c,d}] will plot the graph z=f(x,y) and the curve curve(t) obtained by restricting f to lie on the line through the point {x0,y0} that has been rotated theta degrees with respect to the positive x-axis. If f is restricted to lie along a line parallel to the x-axis then theta = 0, and if f is restricted to lie along a line parallel to the y-axis then theta will be $\pi/2$. (This is not a standard command.)

Together

Together[expression] combines the terms of expression by adding them over a common denominator, simplifying, and displaying the result as a fraction with a single denominator.

ViewPoint

ViewPoint is an option used to change the angle from which a three-dimensional graphics object is viewed. Two of the more common settings are ViewPoint->{0,-2,0}, which views the graph directly from the front, ViewPoint->{-2,-2,0}, which views the graph from the left-hand corner.

VolumeElement

VolumeElement[system] yields the volume element of the coordinate system system. The command VolumeElement is contained in the package VectorAnalysis.m, located in the Calculus folder, and is loaded by Entering <<Calculus`VectorAnalysis`.

Index